John B. Roberts

Paracentesis of the Pericardium

A consideration of the surgical treatment of pericardial effusions

John B. Roberts

Paracentesis of the Pericardium
A consideration of the surgical treatment of pericardial effusions

ISBN/EAN: 9783337270858

Printed in Europe, USA, Canada, Australia, Japan

Cover: Foto ©berggeist007 / pixelio.de

More available books at **www.hansebooks.com**

PARACENTESIS

OF

THE PERICARDIUM.

PARACENTESIS

OF

THE PERICARDIUM.

A CONSIDERATION OF THE

SURGICAL TREATMENT

OF

PERICARDIAL EFFUSIONS.

BY

JOHN B. ROBERTS, A.M., M.D.,

LECTURER ON ANATOMY IN THE PHILADELPHIA SCHOOL OF ANATOMY; DEMONSTRATOR OF ANATOMY IN THE PHILADELPHIA DENTAL COLLEGE; FELLOW OF THE PHILADELPHIA ACADEMY OF SURGERY.

WITH ILLUSTRATIONS.

PHILADELPHIA:

J. B. LIPPINCOTT & CO.

LONDON: 16 SOUTHAMPTON STREET, COVENT GARDEN.

1880.

PREFACE.

SEVERAL years ago, while resident physician and surgeon in one of the hospitals of Philadelphia, I endeavored to occupy my spare hours profitably in the library of the institution. Among the subjects that attracted my attention was that of tapping the pericardium. Although I found recorded a number of isolated instances of the operation, there were few general deductions made, and there existed no complete collection of cases from which statistical information could be obtained. I accordingly published a short journal article on the subject. Since that period my interest in the operation has not abated, and I have read with avidity whatever has come to my notice on the subject; moreover, the opportunities furnished in the dead-rooms of the Pennsylvania Hospital and of the Philadelphia School of Anatomy have enabled me to study the anatomical relations of the parts concerned. There is, so far as known to me, no volume published, and very little contained in the text-books, on this subject; hence I have ventured to offer to the profession this monograph on Paracentesis of the Pericardium, hoping that, as it has no rival, its deficiencies may not be apparent.

J. B. R.

1118 ARCH STREET, November, 1879.

CONTENTS.

CHAPTER I.

CHAPTER II.

CHAPTER III.

CHAPTER IV.

CHAPTER V.

CHAPTER VI.

PARACENTESIS OF THE PERICARDIUM.

CHAPTER I.

THE CAUSES OF PERICARDIAL EFFUSIONS.

In order to discuss properly any method of treating disease, it becomes necessary to direct attention at first to the etiology, symptomatology, and diagnosis of the affection in question. It is therefore proper to devote at first some time to the consideration of the conditions that give rise to pericardial effusions. Every one has seen cases where the autopsy revealed a few drachms of serum in the pericardial sac without there having been any cardiac trouble. This fluid is a transudation that has occurred just at the time of death, or subsequent to that event, coming from the coronary and pericardial veins, and possibly the heart itself, but has not been present as a distinct ante-mortem condition. This must not be confounded with pathological changes, but is to be regarded as a post-mortem appearance, even though at times as much as three ounces* of serum is found. When death has occurred from disease involving great venous stasis of the coronary vessels, the quantity of pericardial fluid found at the autopsy is

---

* Ziemssen's Cyclopædia of Medicine, vi. p. 665.

more apt to be great; in which cases it becomes essential
that the examiner discriminate between effusions taking
place before and after dissolution.

The two great causes of pericardial effusions happening
during life are of course inflammation of the serous membrane, and those conditions which lead to a transudation
of the blood elements into the cellular tissue, and into the
various cavities of the body. The former will furnish
plastic matter and serum in varying proportions according to the character of the inflammatory action, while the
latter will give rise to collections of serum or blood, either
separately or mingled. In the former case there will at
times be a certain amount of coloring matter tinging the
fluid, because of the rupture of some small capillary
vessels. In either case the effusion may become purulent
in character secondarily, but rarely, if ever, is the effusion
purulent at the beginning. Under these two heads there
are many causes which are operative in inducing collections of fluid in the pericardium. Thus the pericarditis
may be idiopathic, which, however, is doubted by many,
it may be traumatic, or it may occur as a secondary affection. Secondary pericarditis is frequently due to acute
articular rheumatism, chronic nephritis, pyæmia, scarla-
tina, and other exanthematous diseases. It may, also, as
is readily appreciated, be the result of inflammation of
the pleura or lungs, which structures, on account of their
proximity to the pericardium, may act as the starting-point
of the process giving rise to the pericardial effusion. In
"Ziemssen's Cyclopædia of Medicine," the occurrence of
pneumonia and pericarditis together is dwelt upon at considerable length.

Pericardial effusion is said to occur only in rare instances after wounds of the sac and heart; whether this is
a true statement of the case, or whether it depends upon

the circumstance that patients with such injuries are apt to die early, I am not prepared to say.

Under the second class of conditions giving rise to effusions into the pericardial sac may be mentioned those diseases, which cause dropsy of the other serous cavities from transudation of serum, such as hepatic, renal, and perhaps even cardiac affections. Pathological blood changes also may give rise to effusion, as is seen in cases of purpura and scurvy, where hemorrhage takes place and hæmo-pericardium results. Such were Kyber's cases.

The result in most instances of pericarditis is that absorption of the fluid occurs, and the patient suffers little subsequent trouble, except there be considerable adhesion of parietal and visceral layer. Hence it is that but rarely does the effusion of rheumatic pericarditis increase to an amount sufficient to call for operative treatment. It does happen, occasionally, however, that the inflammation becomes chronic, the effusion augmented, and the sac more and more distended, until death results from mechanical causes. Such cases must certainly be most amenable to cure by relieving the distended pericardium, and the statistics of the operation of tapping show this to be the fact. The inflammatory fluids furnished in idiopathic, traumatic, and other cases where there is a tendency to recovery, as far as the lesion is concerned, will probably be absorbed in a similar manner. The effusions due to profound blood alteration, and to a general dropsical or anasarcous condition, must be understood as belonging to a different category of causation. I should prefer to attribute these to a process of filtration, by which the blood elements escape, than to a determinate variety of inflammation, though some may feel inclined to say there is always some pericardial inflammation present even from the first.

Intra-thoracic growths may be instrumental in producing effusion, either by instituting inflammatory processes by reason of their contiguity to the pericardium, or on account of emptying their contents into the sac, or by interference with the neighboring circulation due to pressure. Under this head should be placed cases of hydatid and cancerous tumors of the mediastinum, which have been known to implicate the pericardium. Hydro-pericardium has also been known to occur from mechanical causes, such as pressure of an aneurism, disease of the cardiac veins, their obstruction by a clot, or from sudden extreme pneumothorax.

So, again, an abscess or a collection of pus in the pleural cavity may find its way into the pericardium by erosion of the wall of the sac. A very interesting case was seen by me some years ago, where a patch of disease in the aorta allowed the wall to give way, so that the blood escaped somewhat slowly into the pericardium. This case proved fatal of course, and was only understood after the autopsy had been made, when a large amount of blood was found in the pericardial cavity.

THE VARIETIES OF FLUIDS EFFUSED.

The character of the fluid contained in the pericardium must vary with the cause which induces its presence. In general dropsy due to hepatic or renal disease one expects to find serum alone, but if the effusion owe its origin to inflammation, there will be more or less fibrin intermingled with the serous fluid contained in the cavity. This is at times very small in quantity, while in other instances the inflammatory process is plastic in character and very little serum is found, though the layers of pericardium are rendered rough and shaggy, with an abundant deposit of inflammatory lymph upon the whole of their surfaces. With such cases we have nothing to do in the present treatise. Sometimes there may be a small quantity of blood contained in this complex fluid, due to rupture of capillaries, which have perhaps been developed in the adhesions formed during the early stages or days of the pericarditis.

In certain instances of purpura and scorbutus, blood is exuded here as in other situations, and the pericardial cavity becomes distended with pure blood. This is not very frequently observed even in these blood diseases, though Bauer states* that he has seen it happen in otherwise healthy people, especially in chronic pericarditis and where the patients were subjects of chronic alcoholism. Cancer may cause hemorrhagic pericarditis, but the condition is very rare. Many cases of so-called hemorrhagic pericarditis are only instances of blood-stained serous effusion. If an adjacent abscess bursts into the pericardial cavity, or a hydatid cyst is discharged in that direction, the contents discovered when the sac is examined will partake of the characteristics of the fluids belonging to these respective conditions. Pus may result secondarily from any of these effusions, but it is not likely that purulent pericarditis ever occurs as a primary condition. That admission of air to the interior of the sac will induce suppuration is not to be questioned. In Chairou's case, the autopsy showed the existence of pus, though at the time of operation the fluid withdrawn was serous. It has been stated that the pus is not apt to find its way through the overlying tissues to the exterior of the body, but analogy would certainly lead us to expect that purulent pericarditis would evacuate itself in a way similar to that occurring in purulent collections in the cavity of the pleura. It is probable that the infrequency of this pathological sequence in

---

* Ziemssen's Cyclopædia, vi. p. 563.

the former disease is due to the infrequency of purulent pericarditis as an affection, and the liability of death to take place in its early stages. Wyss is said* to have recorded an instance where a rib was worn away and a fistula established which remained patulous for many years. The discharge might take place into the pleural cavity, bronchi, or œsophagus, if the patient survived long enough. At least there seems no pathological reason to the contrary. It is possible that both hemorrhagic and purulent exudations may become absorbed, though this result in the latter case is undoubtedly to be regarded as very exceptional.

QUANTITY OF FLUID.

The amount of the pericardial fluid is a matter of major importance, for it is owing to this factor that there comes in certain cases a question of operation. In some instances there is only sufficient exudation to cause the two serous layers to adhere slightly. This has been called dry pericarditis, but with such cases we have nothing to do. When the inflammation has existed for some time the serous effusion increases with greater or less rapidity; and, if the case become one of chronic pericarditis, where there is no tendency of the absorbents to carry off the inflammatory products, the sac may become enormously distended until the lungs are pushed backwards and laterally, and the diaphragm depressed until the abdominal viscera are actually displaced. As stated above, it is usual to find a small amount of serum in the pericardium as the result of post-mortem changes, but in occasional instances of inflammation or dropsy the quantity is really astonishing. Corvisart mentions an instance where eight pounds of serum were found. Andral gives† a case where two

---

* Ziemssen, vi. p. 564.

† Clinique Méd., 2d ed., 1829, vol. i., observ. iii. p. 15. Quoted by Hayden.

pounds of blood were discovered in this situation. Recently, Dr. Alonzo Clark, of New York, has related* the history of a patient, in whose pericardium was contained one gallon of sero-purulent fluid. Viry found at the autopsy of his case (No. 55) that the capacity of the sac was two or three litres, and that there was on each side a sort of cul-de-sac, which was on a lower plane than the central part of the pericardial cavity. In cases of this kind it would be difficult to get the fluid to flow from the trocar. A case of purulent pericarditis is recorded,† where the tension of the distended sac was so great that a puncture, made at the post-mortem examination, caused the pus to spurt up to the ceiling. In this patient there was an accompanying empyema. Dieulafoy states that the pericardium of a well-grown adult can contain one thousand to twelve hundred grammes of water, and that the sac when injected overlaps the left edge of the sternum from seven to twelve centimetres. The pericardium of an adult male with a normal-sized heart is capable, according to Sibson,‡ of holding from fourteen to twenty-two ounces of water, while that of a boy of six to nine years can contain about six ounces.

Under the influence of the pressure exerted by the quantities observed, as mentioned above, the pericardial sac becomes greatly distended and at times thinned, though the irritation is more apt to give rise to such inflammatory proliferation that the walls are thickened. When the effusion is great, it is usual to find at the autopsy that the heart lies at the back and upper part of the sac, as would be supposed from the attachments of the organ, and the fact that it is heavier than the fluid.

---

* Phila. Med. Times, Nov. 9, 1878, p. 60.

† Lancet, 1863, vol. ii. p. 160.

‡ Reynolds's System of Medicine, vol. iv. p. 305.

In a surgical view the discussion of the amount of effusion deserves considerable attention, for as no one would think of tapping when the fluid is small in quantity, so, on the other hand, no good surgeon should hesitate to tap promptly when the amount of serum or pus is great and threatening the existence of the individual. It is to be understood, however, that the absolute quantity would not be a certain guide even if there was any method of calculating it. The parts can and will accommodate themselves to a large exudation of slow formation, while a rapid accumulation, though comparatively small, will induce the most urgent symptoms. To rather large effusions of serum in this situation the term hydrops pericardii or hydro-pericardium has been applied, and this term, according to Corvisart, is to be used respecting cases where the amount exceeds six ounces. The nomenclature I should prefer would differ in this respect: that, leaving quantity out of consideration, I should call all effusions hydro-pericardium in which the fluid is the result of unbalanced circulation allowing transudation, such as occurs in obstruction of the venous circulation from any cause. To cases of inflammation, of subacute or chronic kind, where there is fluid thrown out, the term pericarditis, with some descriptive adjective, is certainly preferable, and withal more scientific than hydro-pericardium. In a similar manner if blood were extravasated into the pericardium by a simple osmosis induced by profound blood alterations, the term hæmo-pericardium would be proper, whereas, if the blood were furnished by rupture of vessels in a state of inflammation, hemorrhagic pericarditis would be better. According to this nomenclature, the appellation traumatic hydro-pericardium would not be admissible, but traumatic pericarditis with effusion would be legitimate.

CHAPTER II.

THE SYMPTOMS OF PERICARDIAL EFFUSION.

The symptoms of the condition must next claim considerable attention, because the latent character of diseases affecting the internal organs is at times sufficient to throw the attendant off the track. Again, the early recognition of symptoms, pointing to involvement of the pericardium as a sequence of other affections, will possibly enable us to institute a line of treatment to preclude the occurrence of large pericardial exudation. It may be stated in advance that the symptoms alone of pericardial effusion are not sufficiently definite to permit a diagnosis to be made without the aid of physical examination. In fact, it was this circumstance that deterred the earlier surgeons from attempting operative measures. The case did not warrant the performance of an operation until the existence of fluid was determined; but the presence of fluid could seldom be more than surmised until a knowledge of the methods of physical exploration was possessed by the medical world.

In pericarditis there is at times of course some subjective evidence of trouble before the sero-fibrinous exudation occurs; but in true hydro-pericardium, as we employ the term, the effusion is the first step in the history of the condition. After the advent of effusion, the symptoms and signs are similar without reference to etiology.

The symptoms of pericarditis are, as a rule, obscured by the disease which has affected the patient primarily;

hence, as idiopathic pericarditis is rare, it is difficult to present the incipient symptoms to the reader in a categorical manner. There is, frequently, a slight amount of pain or uneasiness experienced, referred to the præcordial region, or shooting in various directions from it; there may be tenderness, especially when pressure is made upon the epigastrium upwards in the direction of the heart. This, however, does not furnish very valuable evidence, for acute gastritis, pleuritis, and hepatic trouble will present a similar symptom. It has been suggested that the study of the special pain due to pericarditis has been perhaps too much neglected, and attempts have been made to study the characters and semeiological value of the pain connected with pericarditis itself, omitting that dependent upon complications.* Absence of pain is said by some to be more frequent in complicated than in simple pericarditis, but this point is of little practical value.

The slight exacerbation of fever and the occurrence of rigors that would be expected as an accompaniment of inflammation of this serous membrane may be absent, or veiled by the febrile action belonging to the primary rheumatic or nephritic disorder. There may be disturbance of the cardiac action, as shown by palpitation and increased frequency of action, which also causes rather hurried respiratory efforts. When the sac becomes occupied by a quantity of fluid sufficient to induce symptoms, we have the same train in pericarditis with exudation, and in hydro-pericardium. The heart and the adjacent organs become involved by reason of the mechanical results of the fluid accumulation. The pulse may be feeble, frequent, and even irregular; there may be more or less urgent

---

* Lo Sperimentale, tomo xl. (1877) p. 419.

dyspnœa or even orthopnœa; and oppression referred to the cardiac region may be present.

Traube has stated that sometimes in copious effusion the left
carotid and radial arteries pulsate
less strongly than on the right
side, but he is unable to explain
it. The disturbed circulation
and respiration is due partly to
the fact that the fluid prevents
perfect dilatation of the auricles,
and hence the venous return
from the lungs is interfered with.
Moreover, there is pressure exerted upon the lungs and left
bronchus. (Fig. 1.) Occasionally
there is fulness of the veins of
the neck, and at times venous
pulsation, due, probably, to the
pressure of the fluid upon the
thin-walled right auricle and the
intra-pericardial portion of the
descending vena cava.

FIG. 1.

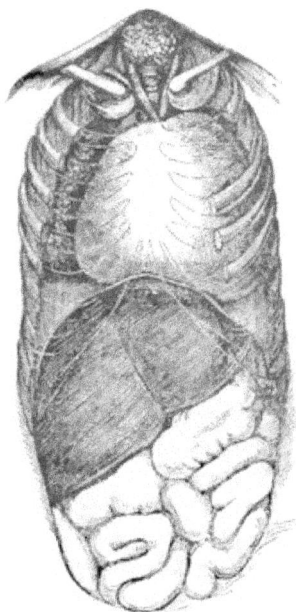

Case of chronic pericarditis in which
three and one-fourth pounds of fluid
were contained in the sac.—Reynolds.

A dry cough, due to reflex irrita-
tion of the larynx, and vomiting are occasional accompaniments. Venous congestion of the right side of the
heart and the lungs, with pallor and perhaps cyanosis of
the peripheral parts of the body, would be expected.
These symptoms are due to the interference with the cardiac impulse, as well as to the pressure of the enlarged sac
upon the veins and lungs, which have no opportunity for
full expansion. Singultus at times occurs, due perhaps
to inflammation of the phrenic nerve; there may be dys-
phagia also as an accompaniment. During this time ner-
vous symptoms, due to imperfect cerebral circulation, are

shown, and the patient may have frequent attacks of syncope, during which he may succumb. If he do not die from sudden syncope on exertion, the end may occur from œdema of the lungs; or, after presenting delirium, he may sink into a comatose condition. The effusion is sometimes complicated with myocarditis, which of course increases the severity of the circulatory symptoms. The equivocal character of these subjective symptoms serves to render prominent the value of the physical signs of pericardial effusion, without a knowledge of which operative measures would seldom have been undertaken. By these we are enabled to watch the course of the affection almost from the moment of hyperæmia, and estimate in a somewhat accurate manner the amount and quality of the contained liquid.

THE PHYSICAL SIGNS.

Let us, then, discuss the results found upon physical exploration in cases of suspected pericardial effusion. In the first place, What is discovered by inspection and palpation? If the pericardium is greatly distended there will be some evidence of increased prominence of the chest wall, with perhaps elevation of the nipple, and it would be natural to imagine that the increase of the left side would be especially marked in the præcordial region. It is stated that the ribs are elevated, thus presenting an appearance similar to what is seen during the time of inspiration. It must be borne in mind that the bulging is seldom sufficient to be a marked feature of the case, since most patients will recover or the disease prove fatal before sufficient fluid has collected to give rise to any great increase in girth.* There have been instances, I believe,

---

* In young children with their flexible parietes this prominence of the cardiac region is of more value than in the old, where rigidity, deformity

where the bulging is said to have resembled the pointing of an abscess. Should there be old adhesions of the pleura, or very rigid chest walls, this sign, as mentioned above, would not be of much value. Again, other affections of the heart itself, such as hypertrophy, and pulmonary affections, or mediastinal growths may present a similar appearance. Fulness of the epigastrium may occur in pericardial dropsy from depression of the diaphragm, and there may also be evident some congestive swelling of the liver. It has been thought that a visible undulation, imagined to be due to the heart's contraction and dilatation setting the serum in vibration, is an evidence of effusion; but this has been denied, and the undulatory movement has, on the contrary, been attributed to the actual cardiac impulse, rendered visible by transmission to the surface, when there is no large amount of effusion present.

Palpation in pericarditis furnishes little information except as to the locality of the apex beat, which can perhaps be determined better by auscultation, provided there is not sufficient fluid to mask all sounds entirely, in which case both palpation and inspection are unavailable. The character of the apex beat may be determined, but it is the change of position to which most importance is to be attached. When the sac is distended the heart can move more readily, and hence, if the patient lie upon the left side, the beat is felt rather more to the left than before the alteration of decubitus. It should be remembered, however, that a certain amount of mobility of this nature occurs normally. Again, in inflammation the effusion is apt to begin about the root of the organ, and consequently

---

from injury, emphysema, intra-thoracic tumor, and other affections of advancing life, serve to alter the physical conformation of the chest.

may push the heart's base downward, and thus, making it take a more horizontal position, thrust the apex beat towards the left, whatever be the patient's position.   When there is sufficient fluid to push the heart backward and prevent contact with the thoracic parietes, the impulse will become imperceptible.   This will vary as to the time of its occurrence with the strength of the heart as well as with the amount of fluid; evidently, more intervening fluid will be requisite to deaden a strong cardiac impulse than a weak one.   Provided the power of the patient's apex beat be known to the observer, it is possible to gain a partial idea of the variations occurring in the amount of fluid contained in the pericardium.   In cases that, from preceding inflammation, have the parietal and visceral layers of pericardium adherent in places, there will be little or no change in the locality or power of the apex beat. Sometimes the impulse can be more certainly felt and appreciated by allowing the subject to sit up or lean forward, since this permits the organ to fall forward towards the sternum, and impinge once more against the costal cartilages and ribs.   Finally, palpation at times furnishes evidence of pericarditis, by reason of a friction fremitus being perceptible to the surgeon as he lays his hand over the præcordium.

Of all the methods of exploration the most useful in the physical examination of suspected pericardial effusion is certainly percussion ; for, by the deviation from the normal area of dulness, and by the variations occurring from day to day, a quite definite idea of the pathological changes can be obtained.   It is impossible to state how small a quantity of serum will make its presence distinguishable by increasing the cardiac dulness.   Bauer says we must not expect notable changes if the quantity be less than one hundred cubic centimetres, though there may be many circumstances to invalidate any conclusions based upon this statement. The same writer is of the opinion that stress is to be laid upon the fact that the effusion increases the area of relative cardiac dulness; meaning by relative dulness the impaired resonance found at the periphery of the region of absolute dulness, where the lung overlies the front of the heart. It is said that there may be considerable serosity in the sac without any change in the area of absolute cardiac dulness, and consequently the change in the region of impaired resonance is worthy of being noted. This statement I believe to be correct, for I have seen in the dead-room the edge of the lung overlie the pericardium to such an extent that there was little or none of its surface exposed. Consequently an effusion of considerable amount could be present in such a case and hardly afford any region whatever of absolute dulness; and if it was found in a case of suspected effusion that there was a small area of absolute dulness, it would undoubtedly be attributed to the ordinary exposed portion of the heart. Hence it is of major importance to define the region of relative or partial dulness, rather than to pay attention exclusively to the extent of absolute impairment of resonance. Rotch says that by percussion of the fifth right intercostal space he was able to diagnosticate the existence of seventy to eighty cubic centimetres of fluid introduced into the sac experimentally.*

The various factors which produce greater intensity, as well as greater extent of the relative dulness, must accordingly be discussed. In the first place, then, the distended pericardium compresses the lung in front of it against the chest wall, and causes impaired resonance by preventing perfect expansion during inspiration, and by furnishing

---

* Boston Med. and Surg. Journ., Oct. 3, 1878, p. 423.

a non-resonant collection of fluid behind the thin layer
of pulmonary tissue. If the lung be non-adherent to the
adjacent tissues, it may be forced laterally and backwards,
and so increase the surface over which there is found a
percussion note of absolute dulness. On the other hand,
if there be old pleuritic bands, or if there exist intra-
pericardial adhesions, the area of dulness will be exceed-
ingly irregular in outline. The comparative elasticity of
the pulmonary tissue has also some agency in the production of a large area of relative dulness, since, if the lung
be more or less non-compressible, it will require greater
intra-pericardial tension to effect its compression against
the anterior wall of the chest. Thus much for the mechanical occurrence of increased dulness in effusions
within the pericardium.

The diagnosis of the condition is rendered much easier,
as may readily be seen, when
the increase in dulness is suddenly developed under the observation of the physician, as
this would preclude the possibility of the variation in percussion being due to cardiac
hypertrophy, chronic indura-
tion of the lung, or any analo-
gous pathological change. It
has long been asserted that the
shape of the region of cardiac
dulness in cases of effusion is
that of a rude, truncated triangle, with the apex at the
root of the great vessels, and
the base downwards. (Fig. 2.)

Fig. 2.

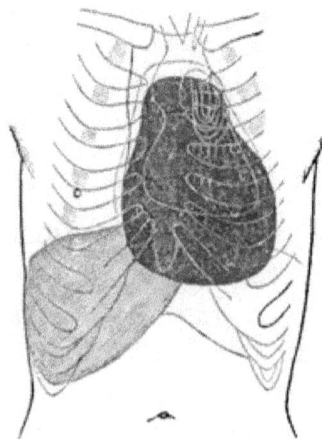

Rheumatic pericarditis, showing region of dulness on percussion at period
of the acme.—Reynolds.

Whether this is true may be discussed subsequently.   One

of the surest signs of fluid in the sac is the existence of relative dulness to the left of the apex, due to the serum pressing the loose fibro-serous membrane away from the heart. Hayden thinks this test not available, because we cannot tell where the apex is normally located.* An important point would be dulness below the apex beat, if that fact were determinable, as it sometimes is. The fluid, if considerable, would tend to depress the diaphragmatic portion of the pericardium away from the heart, which, though heavy, is suspended in the pericardial cavity by means of the vessels at its base, hence the apex beat is found above the lowest limit of dulness. It is usually taught in the class-room that the area of dulness in pericardial effusion is triangular, and this is given as a valuable diagnostic point. If we could always have an otherwise normal pericardium equably distended, without any intra-pericardial attachments, and a pulmonary tissue with no pleural adhesions, and with the normal conformation of outline, we might occasionally see a typical pyramidal, or triangular area of dulness when the patient was upright. That these conditions are almost unattainable in cases of chronic disease, such as gives rise to large effusions, makes it evident that he, who expects to find the beautifully triangular outline described in books and lectures, will be doomed to disappointment.

After I had written the above paragraph, and in fact the greater portion of this little work, and had stated elsewhere my doubt as to the pyramidal area of dulness to be found in pericardial effusion, I fortunately saw the article of Dr. Rotch, of Boston,† containing his elaborate investigations of this very subject. My opinion, which was

---

* Diseases of Heart and Aorta, American Ed., i. p. 388.
† Boston Med. and Surg. Journ., 1878, vol. xcix. pp. 389 and 421.

founded on theoretical considerations, that the fluid would collect at the bottom of the sac and not at the top, as stated by many writers, is the same as that held by him after careful experiments. The results obtained by this gentleman are so interesting that I shall give a summary of his method of experimentation, and an abstract of the practical points obtained. He injected the pericardial sacs of a number of subjects with melted cocoa-butter, and marked out with ink the area of flatness (absolute dulness). Great care was exercised to have the subject semi-recumbent, and to see that the lungs were properly inflated. Subsequently the sternum was removed, and the condition of the intra-thoracic organs carefully noted.

When he injected a small amount of fluid into the sac "percussion gave an increase of præcordial flatness, as follows: beginning at the sixth rib, about two centimetres to the right of the sternum, it passed upwards in a curved line with the convexity outwards to the fourth right costal cartilage at its lower edge, then across the sternum to the upper border of the fourth left costal cartilage, and outwards and downwards to, and to the outside of, the nipple, passing down to the sixth or seventh rib." There was, he says, no vertical increase of flatness, which corresponds with Sibson's statements. It must be recollected that he uses the term flatness to signify absolute dulness. When a large amount of fluid was introduced, the area of absolute dulness was found to be as represented in the diagram, drawn directly from the cadaver (Fig. 3).

The size of the dull area varies with the amount of effusion and the compressibility of the adjacent tissues, and also depends to some extent upon the position of the patient. In the supine position the area will probably be smaller than when the patient is sitting upright or leaning forward, so there may be considerable deviation in the

position of the region of impaired resonance as the invalid
lies upon one or other side of the body. The left border
is apt to be obliquely to the left and downwards; the right
side of the triangle is said to be more vertical than the left,
and follows the border of the right lung near the right edge

FIG. 3.

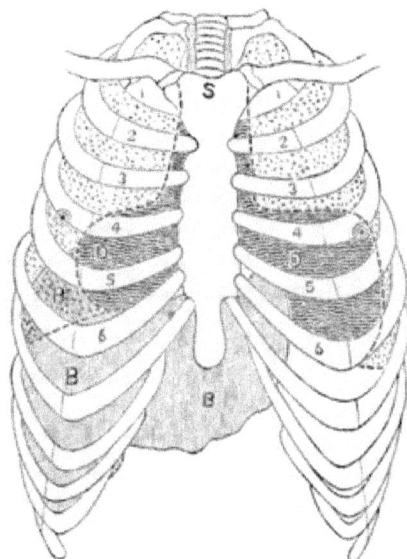

Large amount of liquid introduced into sac.—Rotch.

B, Liver. B′, Portion of liver covered by right lung. D, Area of percussion flatness caused by large effusion. S, Sternum. - - - - - broken line,—Border of lung.

of the sternum. The base of the dull area is in the neighborhood of the sixth or seventh rib. In cases of very large
chronic effusions, a great portion of the front of the thorax
may be dull on percussion. In Pepper's patient, the dull
area extended from a point one inch to the right of the
sternum to two inches to the left of the line of the left
nipple; the upper limit was at the second interspace, and
the base at the level of the seventh rib.

Percussion in cases of large pericardial effusion will

generally show impaired resonance over the upper part of
the left lung; the note is higher in pitch and more tympa-
nitic than on the right side, as in pleural effusions, which
impart a semi-tympanitic note to the subclavicular region.

What will auscultation reveal in cases of pericardial
effusion? The friction sound, due to the rubbing together of the inflamed surfaces of the visceral and parietal
layers, is pathognomonic of pericarditis, but may have
disappeared before the patient comes under observation,
on account of rapid effusion and separation of the surfaces;
or may not appear at all in cases where the effusion pos-
sesses more the character of a simple dropsical transudation,
and where very little inflammatory lymph is furnished.
The presence of a distinct friction sound must not lead us
to believe that there is but little fluid in the sac, for it has
been heard where two pints of fluid occupied the pericardial cavity. It is usually, but not always, heard best near
the origin of the great vessels, and at times becomes more
audible if the body is bent forwards. The valvular sounds
of the heart will as a rule become more feeble or entirely
absent during the occurrence of great effusion, being in-
fluenced very much as the apex beat is modified by the
same pathological condition. The second sound, however,
is usually still heard over the base of the heart and at the
top of the sternum. If there be intra-pericardial adhesions, the heart-sounds may remain quite audible at certain
positions, though the effusion be large.

We have thus seen that the physical signs of effusion
into the pericardium are much more reliable and definite
than any of the symptoms mentioned, and that the diagnosis must be based upon the former, since the latter are
not trustworthy guides to the pathological condition.

THE DIAGNOSIS.

After a consideration of the physical signs of pericardial effusion, it is necessary to show how the attendant may make the diagnosis between this condition and the several diseases that, upon exploration, furnish somewhat similar results. Ordinary cases of pleurisy are easily discriminated by the position of the area of dulness and the normal condition and position of the heart-sounds and apex beat; but sacculated pleuritic effusions in the anterior thoracic region may give rise to much difficulty. Accurate percussion and careful auscultation would probably prove that the heart was displaced to the right in sacculated pleural effusion of the left side. A pleural effusion in the region anterior to the heart pressing this organ backward is conceivable, and as the triangular pericardial dulness is not reliable, such a case would be very perplexing. Retraction of the lung, exposing more than usual of the heart's surface or pneumonic consolidation of the edge overlying the heart, could only be diagnosticated by auscultatory signs, such as normal loudness of heart-sounds and absence of friction in one case, and bronchial respiration in the other. Mediastinal abscesses and growths occurring within the thorax would be shown by irregularity of dulness and displacement of the cardiac sounds. In such conditions the signs would be so variable that it is impossible to lay down any definite diagnostic rules. The exploring-needle or trocar might give information, and in my opinion would be perfectly justifiable in cases where an accurate diagnosis was demanded by the urgency of the phenomena.

That care in physical examination and a correct appreciation of the signs presented by the various forms of intra-thoracic disease is requisite, becomes evident when we recollect that Desault found at an autopsy that he had performed thoracentesis, while the pericardium, which he thought he had tapped, was adherent.   Again, in Béhier's case, a question has been raised whether he did not withdraw the fluid from the pleura instead of the pericardium. That a left-sided pleuritis might put an operator at fault is readily conceivable.

At a meeting of the New York Pathological Society,* some specimens of pericarditis and pleurisy of the left side, taken from a patient whose chest had been aspirated twice, were shown by Dr. Loomis; and yet it was impossible, even after the autopsy, to say whether the needle had really entered the pericardium or not.   The case operated upon by Labric and quoted in the table (No. 34) is even more remarkable.   The child was supposed to have double pleurisy and pericarditis.   He wished to do thoracentesis, and punctured the wall in the fifth interspace about four centimetres outside of the left nipple. More than a litre of purulent serum escaped.   The patient subsequently died, and the autopsy revealed the fact that the fluid had come from the pericardium and not from the pleura, which was adherent to the lung.   This case of unintentional pericardial tapping shows the extent to which the sac may be distended.

Hypertrophy of the heart is usually distinguishable by the heavy first sound, the character of the pulse, and the general symptoms. If a pericardial effusion increases slowly and there is not much diminution in the intensity of the sounds, there will be a slight resemblance to the former condition.   The differential diagnosis among cardiac diseases, which it is most necessary to study thoroughly, is between effusion and dilatation, since in each there occurs

* New York Medical Journal, 1877, vol. ii. p. 634.

extension of the area of dulness, feebleness of the sounds
of the heart, dyspnœa, venous congestion, and syncope.
To show how difficult the diagnosis may at times be, I
give the points which are usually mentioned as serving to
establish the diagnosis: yet an attentive study of them
makes it evident that there is no very pathognomonic sign
to which surgeons can fix their faith. In effusion the percussion dulness is stated to be triangular instead of square
as in dilatation; this is certainly in many instances not the
fact. Again, in effusions a friction sound is often heard
at the base during the period of distended pericardium;
but in hydro-pericardium, or non-inflammatory dropsy,
this must not be expected any more than in dilated heart.
There may be dropsy and venous stagnation, dyspnœa, and
cough in both affections, though perhaps more frequently
and more markedly in dilated heart. And again, the history is of little avail if both diseases have had a slow onset.
The character of the cardiac sounds is perhaps the sign
which oftenest answers a true purpose. In dilated right
heart the sounds are usually clear and sharp, though feeble;
in pericardial effusion they are feeble and distant when
one ausculate at the apex, but more distinct when one listens
at the upper part of the sternum. A sign when the right
auricle is dilated is pulsation of the veins of the neck,
though this may sometimes occur also in effusion. A
careful weighing of the points will usually enable the attendant to establish the diagnosis; but it is to be recollected that a dilated right ventricle has been punctured
by the trocar during attempts to draw off a supposed pericardial effusion. Fortunately, these cases seem to show
that no harm results from this undesirable mistake. Cases
of this kind have come under my notice in medical literature; therefore let one weigh every symptom and sign
in doubtful cases before proceeding to operate.

The diagnostic sign pointed out by Rotch is, if experience proves its accuracy, a valuable aid to the surgeon about to perform paracentesis of the pericardium. He says flatness at from two to three centimetres from the right edge of the sternum, in the fifth intercostal space, would be almost absolutely sufficient to mark the presence of a small or large effusion, unless the opinions of authorities on enlarged heart are proved to be incorrect.* The diagram on page 27, showing the area of flatness in effusion, must be compared with that of enlarged heart (Fig. 4)

FIG. 4.

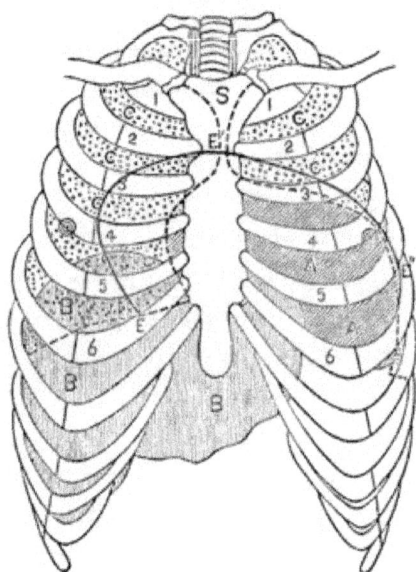

Enlarged heart.—Rotch.

A, Area of percussion *flatness* caused by enlarged heart. B, Liver. B′, Portion of liver covered by right lung. S, Sternum. EE′E″, Line of *relative* dulness of enlarged heart. - - - - - broken line,—Border of lung.

to appreciate fully this assertion. The latter represents the area of flatness due to a heart increased in size by hypertrophy or dilatation, as described by standard authorities,

---

* Boston Med. and Surg. Journ., 1878, vol. xcix. p. 427.

and it is seen that there is little or no absolutely dull area in the fifth right space. He suggests also that puncture can safely be performed in the fifth right space at from four and one-half to five centimetres from the edge of the sternum.

The relation of some of the instances where competent observers were foiled may be mentioned in this connection. Trousseau came very near tapping a case where subsequent post-mortem examination showed cardiac hypertrophy with a very small amount of effusion. Roux's case showed great dilatation of the heart, but no effusion whatever. Roger mentions the case of a child suffering with dilatation and valvular disease in which he had almost decided to do paracentesis, but waited. The child died, and he tapped the chest after death, but then discovered that there was no trace of pericarditis.* Hayden also relates† a case where he was deceived by the physical signs in a somewhat similar manner. The most remarkable instance of an error of this character is that of a woman‡ who had pleuro-pneumonia and signs of large pericardial effusion; as she was almost moribund, a trocar was introduced at the fourth interspace, but to the dismay of the operator dark, venous blood escaped. The instrument was immediately withdrawn, and the patient seemed relieved of the distress and dyspnœa. Four weeks later she died of a complication of disorders, and there was discovered dilatation and valvular disease, but no effusion. The right ventricle had been tapped, and a drachm of blood withdrawn without any shock to the patient. On the contrary, the abstraction of blood seemed to relieve the distended and engorged heart better than phlebotomy,

---

* Bull. del Acad. de Méd., p. 1213.
† Diseases of Heart and Aorta, American ed., i. p. 423.
‡ Transactions Clinical Society of London, viii. p. 169.

as was evinced by the diminution of threatening symptoms and the decreased area of dulness.

Though these histories tend to discourage us in undertaking the operation, still the results of paracentesis in suitable cases are so brilliant, that we must be prepared to proceed if the diagnosis can be made clear. Many patients succumb because tapping has not been performed. Dr. Wilks is quoted* as saying that in three cases of purulent pericarditis, seen by him post-mortem, where tapping would have been feasible, the diagnosis had not been made. Let us, then, endeavor to clear up doubtful points, and make an effort to unravel these knotty diagnostic questions.

At times it may become desirable to determine whether the effusion is the result of inflammation or due to a transudation. The latter has the following characteristics: it is usually a part of a chronic general dropsy, and is apt to be an event subsequent to hydrothorax; there are no severe symptoms, such as pyrexia, at the beginning; there is no friction sound, and very little disturbance of the heart's action; as a rule there is less tendency to a large amount of fluid accumulation, and the area of dulness is more apt to vary in size, and to be altered by changes of posture.

I have not mentioned under the head of diagnosis such affections as cerebral inflammation, gastric irritability, etc., which may be confounded with early pericarditis, because I am discussing the subject in its surgical aspect alone, and hence have only to do with the exudation stage.

THE PROGNOSIS.

The result in cases, which show evidences of pericardial effusion, must vary with the underlying cause of the con-

---

* Lancet, 1872, i. p. 893.

dition, the quantity, and also the quality of the fluid contained in the sac. Inflammatory effusions, such as occur in rheumatism, if of small amount, are to be regarded as favorable in respect to absorption. When the sac contains many ounces of sero-fibrinous exudation, the probability is that the absorbents will be unable to deal with it, and that the sac will become more and more distended, and the fluid in time purulent. In this event the cardiac muscle is liable to become degenerated from the occurrence of myocarditis, which, moreover, is at times a complication of serious import from the early days of the disease.

When pus forms absorption is wellnigh impossible, and the patient is likely to die from exhaustion or the results produced by the purulent collection, unless surgical relief is given by tapping. Occasionally, but very rarely, there may be a spontaneous evacuation of the pus through the chest wall, producing a fistule. In non-inflammatory transudation into the pericardium the prognosis is bad, because it results from some such condition as chronic nephritis, and is but a symptom of serious disease elsewhere. The immediate danger to life in such cases, however, may exist in the pericardial effusion rather than in the primary disease. Moreover, it must be borne in mind that renal symptoms, even the presence of casts in the urine, may be secondary to circulatory disturbance, resulting from the fluid in the pericardium. Pepper's case (No. 57) is an illustration of this important fact.

The hemorrhagic form of effusion is unfavorable, on account of its being due to important blood changes which are inconsistent with rapid restoration of health; still, I should be inclined to consider such cases more amenable to medical treatment alone than those in which pus fills the sac. In the latter event tapping, if the condition can be recognized, should be resorted to at a comparatively early period, before surrounding structures suffer from the pathological processes going on within the pericardial membrane.

CHAPTER III.

TREATMENT.

THE ordinary cases of pericarditis are undoubtedly amenable to treatment by medical means, which should consist of such remedies as are used in the management of pleuritis. In rheumatic pericarditis of moderate severity nothing is required but a continuance of the anti-rheumatic medication. Should the inflammation of the pericardium assume greater severity, it would be necessary to direct our energies to its alleviation, as is also necessary in cases of idiopathic character. When there is hydro-pericardium dependent upon nephritis, or other causal conditions producing dropsy, the attendant is of course to look to the primary affection, and at the same time get rid of the fluid in the sac by making use of absorbents. The activity of treatment depends upon the fact that the anatomico-physiological relation of parts enhances the importance of a very few fluidounces of serum in this locality.

When acute sthenic pericarditis is diagnosticated, it may at times be advantageous to apply leeches or cups to the præcordium, but I have little idea that general blood-letting is often required. Let it be known, however, that I should not hesitate to employ it if the case seemed a proper one. Those who follow the German teaching will most probably resort to cold locally, as an antiphlogistic and anodyne agent, and it would seem to be a rational

measure. I must admit, however, that my preference is
for warm dressings, such as poultices, since it is certainly
a fact that great relief is afforded by jacket-poultices, or indeed anything furnishing heat and moisture, in pulmonary
inflammations; therefore they seem indicated in cardiac
inflammations. Stimulants should be avoided, and small
doses of cardiac depressants would be advantageous in this
acute form, if care were observed to watch the effect.

In order to keep the inflamed organ quiet opium is of
value, if the patient be not suffering from chronic Bright's
disease, when it must be cautiously employed. Aperients,
diaphoretics, and the like come in as adjuvants, as in all
inflammations. As soon as there is evidence of cardiac
failure, whatever be the stage of the disease, digitalis is
indicated, and may be combined with alcohol in some
form. As salicylic acid seems to exert a favorable influence on rheumatism, it may have a tendency to prevent
the occurrence of pericarditis; so also the alkalies and
bromide of ammonium may possibly have a beneficial
effect. When effusion of serum or exudation of lymph
has occurred, our attention is directed to preventing the
increase, and promoting the absorption of these products.
Tincture of iodine, blisters, iodide of potassium, diuretics,
and hydragogues all come into play. At the same time
this action is to be assisted by tonics, stimulants, nutritious
diet, and digitalis; for now the patient has begun to lose
strength, and the vigor of the cardiac muscle is impaired.
Perhaps also myocarditis may have rendered the heart
more feeble than the pericardial inflammation would induce of itself. My own instincts would lead me to rely
upon poultices and anodynes, iodide of potassium, acetate
of potassium and juniper, digitalis and tonics, in the
various stages of the affection, with restriction of the
amount of liquid drank during the time of effusion. If

hemorrhagic pericarditis was suspected, it would be well
to use the preparations of ergot in addition to the regimen
and medication adapted to the scorbutic or purpurous
condition.

I have said nothing regarding mercury as an antiplastic
agent, as its value is doubted by so many recognized
authorities. In the stage of exudation it could perhaps
be advantageously combined with the iodide of potassium
and the diuretic chosen. In cases of sudden asphyxia, relief may be obtained perhaps by venesection, which acts
mechanically by reducing the amount of blood to be oxy-
genated and renders absorption more active.

When the line of medical therapeutics indicated has
been followed, and the effusion nevertheless continues to
augment, until symptoms of gravest import arise, and
there is danger of fatal consequences, or when the pericardial fluid becomes purulent, which at all times means
mischief, we must throw aside medicinal agents and turn
to the domain of surgery.

Whenever dropsy of peritoneum, pleura, ovary, or knee-
joint baffles our medical armamentarium, we have recourse
to operative measures, and decide that that which cannot
be absorbed must be removed by tapping. Is it possible
to apply this same logic to pericardial effusions? and if
so, what are the results of such an operative procedure?
This brings us to the consideration of paracentesis as a
means of relief in pericardial effusions.

HISTORY OF THE OPERATION.

Paracentesis was proposed by Riolan* as far back as
1649, in these words,—" Si non possis exhaurire istud per
hydragogen, licetne terebra sternum aperire intervallo

---

* Enchiridion Anatom., lib. iii. c. 4 (1649).

pollicis a cartilagine xiphoide?" though I can find no instance of its being performed until Romero, of Barcelona,
operated successfully, and reported his cases to the Faculty
of Medicine of Paris. Of the date of these operations I
am in doubt, but they are mentioned by Mérat in his work
published at Paris in 1819.* The operation was not adopted
because of the difficulty of making a differential diagnosis
between pericardial effusions and other affections, and
because of the supposed vulnerability of the heart. That
the first reason was valid is shown by the fact that some
of the earlier cases of so-called tapping of the pericardium
were not instances of this operation at all. Desault mis-
took† a circumscribed pleuritic effusion for pericardial
dropsy, and actually operated, but the subsequent autopsy
revealed the fallacy. Larrey has been reported as performing paracentesis of the pericardium, but this also is probably not the fact.‡ Thus it is seen that the obscurity
involving thoracic diseases before the application of physical examination to the unravelling of their mysteries, was
doubtless the chief cause of rejecting the operation. Van
Swieten recognized the difficulty, but truly says, " Ten-
tandum esse potius anceps remedium quam nullum, dum
certa pernicies imminet,"—A doubtful remedy must be
tried rather than none at all, when death is certainly
threatening. Hence we may readily perceive that the
doubtful remedy would have been tried had the surgeon
been certain that the presence of a large pericardial effusion was the cause of the threatening symptoms. There
undoubtedly lingered also a feeling that the heart was too
vital an organ to allow the rude approach of instruments,
for Mérat informs us that the Faculty of Medicine at Paris

---

* Dictionnaire des Sciences Médicales, xl. p. 372.
† Œuvres Chirurg. recueillies par Bichat, ii. 1798 (Trousseau).
‡ Bulletin des Sciences Médicales, 1810 (Trousseau).

did not allow the report of Romero's successful cases to be printed in their Transactions, lest this most delicate operation should thus be sanctioned.

Notwithstanding the discouraging circumstances surrounding the operation, writers continued to advise it in suitable instances, and operators of sufficient courage began to appear with greater frequency as the profession gained more and more insight into the physical signs of the condition. Karawagen employed the trocar in treating scorbutic pericarditis, and was soon followed by Kyber, in similar cases, who was so fortunate as to have four recoveries subsequent to the operation. The importance of the subject was scarcely appreciated until thoracentesis became established as the proper method of dealing with large and chronic pleural effusions; but even then there was, and up to the present time there remains, a feeling of distrust in respect to the operation. In an article published three years ago* I endeavored to place the operation on a firmer foundation, and trust that I have been able to do something in that direction.

ANATOMY OF THE PARTS CONCERNED.

Before entering upon the consideration of the operation itself, it will be profitable, perhaps, to spend a little time in rehearsing the anatomical relations of the pericardium. The pericardium is a fibro-serous sac enclosing the heart and the roots of the great vessels, or rather, surrounding and being reflected upon them in such a manner that the serous layer serves as a covering to them in the same manner as the peritoneum does to the abdominal organs. The outer portion of the serous layer is reinforced by a fibrous layer. The space between the visceral layer and

* Paracentesis of the Pericardium, N. Y. Medical Journal, Dec. 1876.

the outer fibro-serous layer is the cavity of the pericardium, and it is here that the effusion collects in cases of inflammation or dropsy. The sac is conical, with the apex attached around the vessels about two inches above their origin and directly behind the top of the sternum, and with its base connected with the centre of the diaphragm. The pericardium and heart lie in the middle mediastinum, between the two pleural sacs. The costal pleura is reflected upon the outside of the pericardial sac on each side as it passes backward from the sternum to go to the root of the lung; from the lateral region of the pericardium it passes upon the lung to become continuous with the pulmonary layer. Near the middle of the sternum the two pleuræ are sometimes in contact for a short distance, but they diverge above and below.* The pleural sacs cover the pericardium in front, except a narrow strip running vertically along the left edge of the sternum; hence, unless the aspirating trocar puncture at this point, the pleura must be injured before the pericardium can be pierced, when the parts retain their normal relations.†

The heart, covered by the visceral layer, lies within the fibro-serous layer, behind the lower two-thirds of the sternum, and in an oblique direction, with its base directed towards the right shoulder. It extends about three inches to the left of the median line, but only about one and a half inches to the right of it. The upper border is on a level with the superior edge of the third costal cartilage, while its inferior surface lies upon the diaphragm with the pericardium interposed. The apex of the heart is located between the fifth and six cartilages, about one inch inside and two inches below the left nipple. The lower border

---

* Braune's Topographical Anatomy, Plates XII. and XIII., English edition, 1877. See also p. 106.

† See woodcuts in Braune, on pages 102–113, and Plates XII., XIII.

4

of the heart, in a frozen section made by Braune,* corresponded with the lower edge of the fifth rib, while the pericardium extended about half an inch lower, and contained about a tablespoonful of frozen fluid.

To determine the extent of the normal pericardium, I recently made an examination of an emaciated female subject of moderate stature. She had died of phthisis, but there was no pleural effusion to interfere with the normal relations. I sawed through the sternum between the second and third cartilages, and turned it carefully downwards, so as not to displace the pleuræ or pericardium. Having cut a slit in the pericardial sac, I introduced my finger, and obtained its dimensions by placing the finger at the right and left limit successively, and thrusting a long pin through the overlying tissues until it touched the end of the finger. The measurements were then taken with accuracy. There was about half a fluidounce of fluid in the cavity. The sac extended to the right of the median line of the sternum five and a half centimetres, to the left of the middle of the sternum seven centimetres. The lower border of the pericardium in the middle line corresponded very nearly with the base of the xiphoid appendix, but at its apex the sac descended about one and one-half centimetres lower. This point was nearly as low as the inferior border of the sternal end of the sixth rib, where it unites with the sixth cartilage. Hence a line drawn from a point just above the lower edge of the end of the sixth *rib* to the junction of the xiphoid and gladiolus would correspond with the oblique floor of the pericardium, and give its lowest boundary. It must be remembered that the floor of the pericardium ascends as it passes backward upon the curved upper surface of the dia-

* Op. cit., p. 106.

phragm. This varies during respiration. These measurements are probably approximately correct, though the size of the individual would cause differences. When the sac is distended with a large effusion the diaphragm is pushed down, and the limits of the pericardium increased laterally. The front of the heart and pericardium is separated from the thoracic wall to a great extent by the lungs, though these organs do not cover as much of the pericardium as the pleuræ, which extend farther forward. The right lung comes to the middle of the sternum, while the left, at the level of the fourth cartilage, slopes to the left, leaving the ventricles uncovered by pulmonary tissue. This gives a notch in the edge of the left lung, exposing the right ventricle and the apex which is a part of the left ventricle. This uncovered portion is rudely triangular, and can be indicated by a circle two inches in diameter, with its centre midway between the nipple and the end of the sternum.*

The left pleura in the subject mentioned ran straight down quite near the left edge of the sternum, until, about on a level with the lower end of the gladiolus, it sloped off to the left; but this, as is seen, did not correspond with the retraction which occurs in the border of the lung in the vicinity of the fourth and fifth interspaces. This divergence of the pleura from the sternum occurred, I may say, just about where the pericardium and diaphragm came in contact, and the pleura then passed upon the diaphragm. It must be understood that the measurements spoken of are about the average, and that pathological conditions, and possibly anatomical variations, may cause differences in the mutual relations of the organs. It is easily appreciated that an hypertrophied heart will show different measurements, and that in the event of large pericardial effusion

---

* Holden's Landmarks.

or of pulmonary disease the lung may not intervene between the heart and the ribs. The pericardium may be greatly distended by the fluid, and depress the diaphragm to a considerable extent, and then correspond in level with a lower rib than the one mentioned. As showing what strange anomalies at times occur, I may mention that a case has been recorded* where the pericardium was not attached to the tendon of the diaphragm at all, but was a bag simply resting on that muscle without any continuity of fibres between them.

CASES SUITABLE FOR OPERATION.

The indications for tapping the pericardium are easily formulated. Whenever the effusion, whether it be serum, pus, or blood, accumulates so rapidly or in such quantity that it threatens to destroy life and refuses to undergo absorption by ordinary treatment, it is the duty of the attendant to tap the distended sac. It is an operation which gives immediate relief by removing the cause of the urgent symptoms, and the figures which I shall deduce from the cases collected prove that the operation itself is not a serious one. How long we are to trust to medical means before attempting surgical interference must be left to the surgeon or physician in charge of each case; but I greatly fear the tendency is to wait too long rather than to operate too early. All surgeons see cases of strangulated hernia lost because of valuable time having been squandered in useless attempts at taxis; so patients have died with large pericardial effusions, because the fluid was not pumped out with an aspirator before the hydrostatic pressure upon the heart, perhaps enfeebled by disease, became intolerable, and the central organ of circulation stopped. That

---

* Journal of Anatomy and Physiology, 1871, vol. v. p. 114.

this picture is not overdrawn is seen by the records of post-mortem examinations, where many fluidounces of serum have been found in the distended pericardium. Hayden, in his work on diseases of the heart and aorta, says distinctly, "On two occasions at least I have had reason to regret the omission of its performance."*

Some writers think the operation unsuitable for cases of hydro-pericardium from Bright's disease, and contra-indicated in hæmo-pericardium as well as in purulent effusion from acute osteo-myelitis and pyæmia, because, they say, the local trouble is only a part of the disease, and there is something more which the operation cannot benefit. True enough; but if it is the local effusion which menaces the life of the patient at the present moment, it is the enemy that must be opposed. It is only palliative, but palliation is not to be despised, since it gives time for other measures to be adopted that may reach, at least in part, the more distant and more general disease. Admit that the operation is only palliative, should one hesitate to perform it because the patient may die in a few days or weeks of some concomitant disorder? If paracentesis was likely to cause or hasten the fatal issue, or if it augmented the suffering of the individual while doing him no good, it might be allowable to hesitate; but that such is not the case is seen by intelligent observation in all parts of the world. Who would decline to tap an immensely distended abdomen because the patient suffered at the time from incurable hepatic disease, or to draw the fluid from the pleura because the patient was tuberculous?

There are, undoubtedly, conditions in which paracentesis of the pericardium is followed by much more brilliant results than in others. The most favorable cases are those

---

in which a rapid pericardial effusion has occurred as a complication of rheumatism. Here there exists a disease that usually has a favorable prognosis, and if the fibrino-serous fluid is withdrawn from the pericardium the danger of the heart complication is averted, and the patient is saved from impending dissolution.

If the pericarditis is chronic, or accompanied by pulmonary or pleural disease, there is less chance of obtaining such satisfactory results, though many cases have been reported where return to health, as far as the pericardial affection was concerned, has followed the performance of the operation on a moribund patient. In these instances the failure of perfect convalescence has usually been due to the incurable character of the pulmonary lesion, or to the textural changes that have taken place in and around the pericardium during the continuance of the inflammatory products within the sac. In certain instances this could have been averted by an earlier resort to operation. When the fluid is purulent it will probably be necessary to repeat the tapping, but I cannot see why it would not be advisable in certain cases to establish a continuous drain, as is done so effectively at times in empyema. This matter, however, must be discussed later, though it may be said here that a second tapping does not appear to be any more dangerous than the primary operation. In fact, it might be looked upon as less so, since there is great probability of the pleura having become adherent after the first operation.

In dropsy of the pericardium from Bright's disease, it may be admitted that the fluid is at times absorbed with great rapidity, and that the operation does not affect the primary disease, if it be primary and not secondary to the cardiac trouble; but still I say tap, if there be evidence of failing circulation and respiration, which in the best judgment of the attendant depends on the effusion.

Dr. Pepper's case (No. 57) shows this, for the kidney symptoms were believed, after the lapse of time had permitted observation, to be due to the pericarditis. Before the operation the urine was slightly albuminous, and contained tube-casts, while some weeks subsequently this symptom had entirely disappeared. What a different result would have been obtained if paracentesis had been rejected in this instance, because there was a suspicion of Bright's disease underlying the pericarditis!

In hemorrhagic pericarditis the prognosis becomes very unfavorable, but not hopeless, if we credit the cases of Kyber, Karawagen, and others. Where there is a pyæmic causation the result must be still more doubtful, and yet in even these instances the operation seems to me justifiable, provided the urgent dyspnœa and depression appears to depend on the obstruction to the circulation.

I am inclined to go even a step farther, and suggest the employment of paracentesis in those rare cases of pneumo-hydro-pericarditis which occasionally present themselves for treatment. A most interesting case of this kind has been reported* by Dr. J. F. Meigs, in which I should have been induced, I think, to have performed the operation of tapping. The presence of air and fluid in the pericardium was diagnosticated before death. At the autopsy there were found evidences of pericarditis, with air and eight to twelve fluidounces of reddish-brown liquid in the sac. The air had entered by an opening leading into the œsophagus, due, Dr. Meigs thinks, to "an effort of nature to evacuate the diseased contents of the pericardium, as happens in the case of empyema when the latter is cured by natural processes." As the patient had had double pleurisy and subsequent pericarditis, he thinks

* American Journal of Medical Sciences, January, 1875, p. 94.

this explanation should be accepted rather than the sup-
position that an accidental ulceration of the œsophagus
should have occurred coincidently. If the case is inter-
preted in the former way, it is certainly a strong argument
for the adoption of paracentesis in pericardial effusions;
and then it may be advanced that tapping, even after
pneumo-pericardium had occurred, would have given an
opportunity for closure of the abnormal orifice.

After it has been decided that the operation is a proper
mode of treating the patient under consideration, it is not
justifiable to delay a long time before resorting to it. In
very chronic cases the immediate danger of delay may be
less, but there is constantly going on a succession of pathological changes, such as thickening of the membrane, adhesion of the pericardial tissue to the structures external to
it, chronic pulmonary congestion, and obstruction to the
venous return from the head, due to the hydrostatic pressure on the right auricle and descending vena cava. On
the other hand, undue haste in operating on acute effusions
is to be deprecated, because large effusions at times disap-
pear with great rapidity under medical treatment. Still, it
appears that more harm is liable to result from delay than
from an operation done early, if due care be taken in the
manner of performing the operation. The danger of delay-
ing the removal of the fluid has not only reference to the
possibility of sudden asphyxia, but also to the occurrence of
myocarditis and fatty degeneration of the muscular wall of
the heart. The existence of inflammation or degeneration
of the cardiac muscle is more often found in purulent and
hemorrhagic cases than in others, but it is liable to occur
in all instances. Whether it be due to pressure interfering
with the blood supply of the organ, or to the increased work
thrown upon the heart, matters little, the fact remains that
such lesions do occur as accompaniments of pericarditis.

CHAPTER IV.

METHODS OF OPERATING.

WHEN the operation first suggested itself to observers and writers, the question that immediately arose was as to the method of reaching the distended pericardium. The earlier authorities, such as Senac, Skielderup, and Laennec, proposed trephining the sternum above the xiphoid cartilage; but this is, of course, not to be thought of at the present day, for the operation would be much more serious if complicated with the suppuration that must needs result from such a procedure. Others suggested that a preliminary incision should be made at the point selected, and the tissues divided layer by layer, as in herniotomy, until the pericardium should be reached, when a trocar was to be thrust into it. Others thrust the ordinary hydrocele trocar directly into the sac.

At our day aspiration is the method preferred to all others. In this method there is no danger of osseous suppuration retarding convalescence by the possible burrowing of pus among the thoracic tissues; there is no long incision which might cause hemorrhage from injury to the internal mammary artery or one of its branches; the entire amount of fluid can be drawn out, there is little danger of the pericardial fluid dribbling into the pleural sac if it should happen to be punctured, and the fine needle will do little harm even if it pass through the edge of the lung or strike the ventricular wall itself.

In addition to the advantages mentioned, the entrance

of air is prevented, and thus the liability to the occurrence
of suppurative pericarditis subsequent to tapping is reduced to a minimum, and the wound is such a trivial one
that it amounts to nothing, giving little pain at the time
of operation and soon healing, so that an autopsy, even
if made early, shows only a small ecchymotic spot in the
tissues. The risk of wounding the mammary artery is reduced to a mere chance, because of the diminutive needle
that can be used when suction power is employed.

If a trocar of ordinary kind is used, the thickened and
hard pericardium may be pushed in front, because of its
loose attachment to the thoracic parietes; but the needle
of the aspirator can be thrust gradually through the tissues
without the same probability of displacing the sac walls.
Again, the vacuum attached to the needle shows the surgeon the very moment the sac is entered, because the fluid
at once fills the tube and is seen flowing towards the re-
ceiver. This is a matter of paramount importance, since
otherwise the needle might be pushed onward into the
ventricle.

As we have stated that aspiration is to be adopted
in performing paracentesis pericardii, it may be asked
whether the ordinary sharp-pointed needle or the trocar
and canula, furnished with the aspirator as sold in the
shops, is the better instrument for perforating the pericardial membrane. There is an objection to having the
pointed needle or the sharp-edged canula within the sac
when the fluid has been partly withdrawn, as there is a
possibility of the heart becoming scratched.

There has been described* by Fitch the so-called dome-
shaped trocar, in which a blunt or "round-ended" fenes-
trated canula slides *within* the penetrating needle. (Fig. 5.)

---

* New York Medical Journal, April, 1875.

This, if made small and adapted to the aspirator, would
be an admirable instrument. In the " New York Medical

Fig. 5.

Fig. 6.

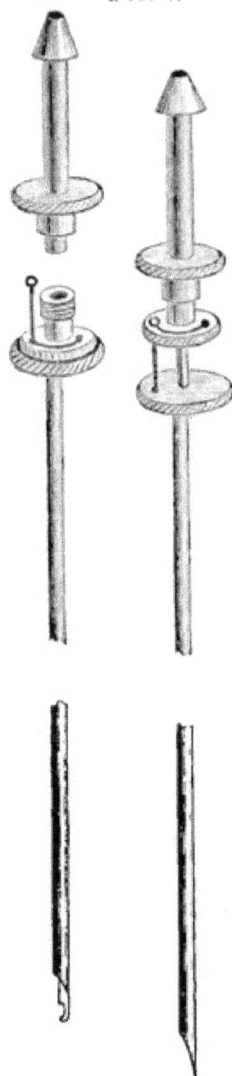

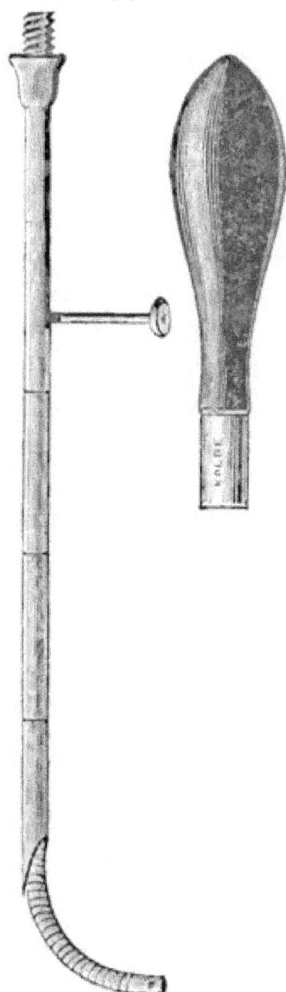

Fitch's trocar.   Roberts's trocar.

Journal" for April, 1877, page 384, I have described an
aspirating trocar which is figured above. (Fig. 6.) It

consists of a small, needle-pointed cylinder, within which
slides, on Fitch's principle, a canula attached to the air-
pump. The canula at the end is made flexible by a spiral,
and when it is thrust out beyond the end of the needle
curves downward; but when it is pulled backward the end
becomes straight, and is entirely concealed within the
outer puncturing needle. The extremity of the canula is
pierced with a hole and there are also two other fenestræ
just above, to give exit to the fluid. The method of using
it is as follows: the canula is drawn back until its flexible
end is hidden within the needle, and the hose from the
pump is attached to a small tube fixed at a right angle to
the posterior extremity of the canula. The outer puncturing instrument is thrust into the integument, and the
operator immediately causes the vacuum to exert suction through the internal canula; consequently a flow of
serum shows the moment that the instrument has entered
the cavity of the pericardium. The surgeon immediately
withdraws the needle a little, and thrusts the internal flexible canula into the sac, so that the point of the trocar is
guarded, and there hangs in the sac a blunt, flexible tube,
against which the heart can strike with impunity. This
aspirating trocar is innocuous to the heart, and the curved
extremity allows suction to be exerted to a considerable
extent at the very bottom of the pericardial cavity. If
the canula becomes plugged with flakes of lymph, the
handle can be unscrewed, the inner portion withdrawn,
and the hose attached to the end of the penetrating needle,
which then acts as a large, ordinary aspirating needle.

Dr. Pepper, after operating on the case reported in
the table, had made a delicate double canula, the inner
tube of which is furnished with a fine needle-point. The
movements of the inner tube are regulated by a button
which moves along a slot in the outer tube. After intro-

duction the inner tube is withdrawn until its point is sheathed.*

When I first described my trocar, mentioned above, I was only acquainted with Fitch's trocars of large size as used in ovarian dropsy, but subsequently I found that he had described small capillary instruments, such as would be required in paracentesis of the pericardium. These, it seems to me, furnish us with the most perfect trocar for the purpose.

The capillary trocars of Southey, invented for draining the serum from œdematous limbs, might, if long enough, be used where it was thought proper to establish a continuous flow of fluid from the pericardium. I merely mention them to the reader because they have been used for paracentesis abdominis,† but do not think the occasion for their use in pericarditis is likely to arise.

In performing the operation, Potain's aspirator with a vacuum jar (Fig. 7) is best, because the vessel can be emptied of air before the trocar is introduced, and there is then no shaking produced by pumping after the pericardium has been perforated. The stop-cock should be turned so as to allow the atmospheric pressure to exert its force, as soon as the fenestra of the trocar is buried beneath the integument. Thus the operator is apprised of the entrance of the trocar into the fluid collection.

If the fluid reaccumulates I can see no great objection to repeating the operation again and again, as was done in Gooch's and Bouchut's cases (Nos. 43, 48). Roger looks upon repeated operations with disfavor,‡ but the evidence deduced from the cases tabulated shows that

* The Medical News and Library, Philadelphia, March, 1878, p. 40.
† Monthly Abstract of Medical Science, November, 1878.
‡ Bull. de l'Académie de Médecine, 1875, p. 1283.

there is no more risk in second than in first tappings, for
in the cases where the operation was repeated there is
no testimony that the procedure did harm.   In many instances the patient subsequently died, I admit, but the
time after operation was on an average too long to attribute the result to the paracentesis; and, moreover, the re-

Fig. 7.

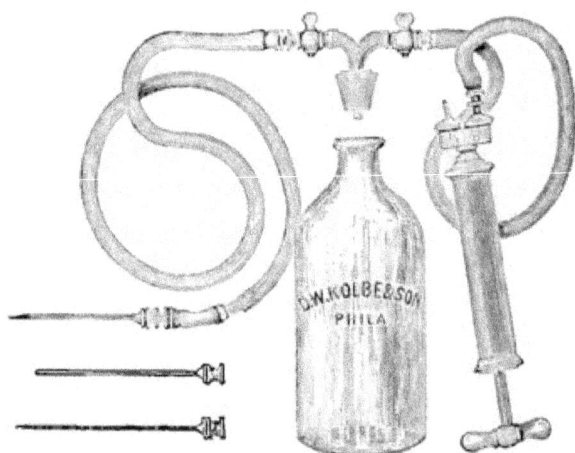

Potain's aspirator.

accumulation was as a rule due to the fact that there was
some complication rendering repeated effusions probable.
Aspiration is repeated in pleuritis and synovitis, and so it
may be in pericarditis if there is a secondary effusion.

Some operators have injected iodine into the sac, after
the withdrawal of the fluid, with the idea of modifying
the secreting surfaces, or of inducing adhesion of the
layers of pericardium.   Aran's successful case (No. 19),
in which tapping and injection was done twice, shows
that this is not improper treatment.   Kyber considered*

---

* Monthly Retrospect of Medical Sciences, Edinburgh, March, 1848.
See also Guenther's Blutigen Operationen, IV. iii.

adhesion the method of cure, because he found it had
occurred in three successful paracenteses where autopsies
were obtained long afterwards; and the fact that Pepper's
case showed close and universal adhesions proves the
probable correctness of the earlier operator's view.

Injection was adopted by Aran as he had found it
satisfactory in pleural effusions, and others have adopted
the procedure. Roger does not think it applicable,* because there is, he believes, danger of suppuration in the
sac, and of effusion of pus into the mediastinum, and because there is not the same tendency as in pleurisy to retraction and approximation of the two layers. My own
opinion is, that since we have the aspirator, instead of the
old trocar, and can prevent the entrance of air, the injection of iodine is not indicated as it was formerly, perhaps,
when the admission of air was liable to change a serous
pericarditis into a purulent one. After simple tapping adhesion will very probably occur, unless the effusion be due
to a passive dropsy, in which case adhesion is not necessary to a cure; hence injections are not required either in
inflammatory effusions, where adhesion will be apt to take
place without any aid, nor in hydro-pericardium, where relief of the condition depends on other factors than adhesion.
The simple withdrawal of the fluid, with repeated operations if reaccumulation take place, is the proper course to
pursue. Should the pericardial contents become purulent,
or if I were afraid of this contingency, then injections of
carbolized water would be preferred to an iodized solu-
tion.

This brings us to the management of purulent pericarditis, which is probably never primary, but secondary to
some other form of effusion. Examination of the cases

---

* Op. cit., p. 1282.

published will show that the operation has at times been done for existing purulent effusion, while at other times the effusion was at first serous, but subsequently became purulent from admission of air, or from some unknown factor, probably due to the paracentesis. How are such cases to be treated? At first there is no doubt that simple tapping is to be resorted to, supplemented possibly with injection and washing out of the cavity with carbolic acid solutions of the strength of one to twenty or thereabouts. If the pus reaccumulate, it is almost impossible for it to become absorbed, and it will increase in quantity and cause pathological alterations in the pericardium and surrounding structures. Hence I would advise a repetition of the operation with carbolized injection somewhat in the manner of Callender's hyperdistention of abscesses, though of course this could not be practised with the amount of pressure justifiable in ordinary situations. If repeated tapping becomes necessary from the rapid secretion of pus causing imminent danger, I can see no reason for objecting to an opening being made to secure permanent drainage. We know the results produced by imprisoned pus; we know that purulent pericarditis practically means death unless it is removed by absorption or operation, to permit of adhesion taking place. Again, we see reported instances where pericardial fistulæ have been established by nature without deleterious results. Wyss has described* a case where a rib was worn away and a fistule established that remained patent until death occurred years afterwards. Villeneuve (see Case 47) had a fistulous track remain for a number of months after paracentesis of the pericardium, but it finally closed.

These facts, coupled with the surgical axiom that pus

---

* Ziemssen's Cyclopædia of Medicine, vol. vi. p. 564.

should always be allowed free egress, even if the abscess
be in so vital an organ as the brain, influences me in my
opinion that purulent pericarditis demands treatment iden-
tical with purulent pleuritis. I have seen such excellent
results follow the introduction of drainage-tubes into the
pleural sac for the treatment of empyema, that I would
recommend a similar line of action in the cases under discussion. Let a drainage-tube be placed in the pericardial
sac, and let the surgeon daily wash out the cavity with
disinfecting solutions. This resort is to be adopted of
course only after tapping has been found unsuccessful. It
seems less dangerous than repeated tapping, because it is
possible that the heart may become adherent to the parietal layer between two successive operations, and sustain
injury from the puncturing instrument.

I am aware that this opinion differs from that of Roger
and of Pepper, and would most probably be looked upon
as hazardous by many. In an article written* in 1876, I
held this opinion, and the subsequent study of the cases of
Villeneuve, Juergensen, and Viry, which I had not then
seen, and other facts bearing on the subject, serve only to
confirm me in this belief. I admit that deductions from
pleural conditions must be received with the understand-
ing that there are two lungs, each of which can supple-
ment the other, while there is only one central organ of
circulation; still the thickening of pericardium, the maceration and disintegration of the cardiac muscle, the liability to spontaneous evacuation, all convince me that the
drainage-tube, or, if you choose, free drainage, is a better
alternative. Whether or not the strict antiseptic dressing
of Lister should be insisted upon I leave to the judgment
of the operator. The object to be attained is the restora-

---

* New York Medical Journal, December, 1876.

tion of the patient to at least a fair degree of health, and
that universal pericardial adhesion is not incompatible
with this is proved by many pathological examinations.

THE POINT OF PUNCTURE.

A survey of the recorded cases of paracentesis of the
pericardium will show that there is a considerable diver-
sity of opinion as to the best point at which the puncture
should be made. The pericardium when distended may
be reached by needles thrust in at several points, as can
be comprehended if reference be made to the anatomical
description of the membrane. There are, however, complications arising from the needle's introduction at certain
points and not at others, that make a careful study of this
question of paramount importance.

When the puncture is made high up there is great liability of wounding the auricle, because the pericardium
cannot be distended at its upper part as it can at the lower.
This accident is to be avoided, since wounds of the auricle
are of grave prognosis, much more so than wounds of the
ventricles, which indeed seem to be of little moment.
Again, if the aspirating instrument is introduced too low
down it may injure the diaphragm, or not enter the pericardium at all. A point selected too far to the left en-
dangers the lung, and if located about a quarter of an
inch from the border of the sternum may place the integ-
rity of the internal mammary artery in jeopardy. Thus
it is evident that a point is to be elected, free from complicating sequences, which gives ready and perfect access
to the pericardial cavity.

Let us first consult what has been said by previous
writers on the subject, and endeavor afterwards to determine which point is to be recommended for its safety,
convenience, and efficiency. Larrey thought the proper

position was to be found between the xiphoid cartilage and
the cartilages of the ribs of the left side;* while Trousseau preferred the fifth or sixth interspace close to the
sternum, and said that if the cartilages were too close to
admit the trocar, a portion of the cartilage might be
trimmed away. A somewhat similar level should be
chosen according to the judgment of Dieulafoy, who
advocates the fourth or fifth interspace, but goes to the
distance of about six centimetres from the left edge of
the sternum.† He selects this position because he found
by experiment that the maximum transverse diameter of
the distended pericardium coincided with the fourth interspace or fifth rib, and that it is here that the notch in the
edge of the left lung is situated; hence the danger of
piercing the lung is reduced to a minimum. The point
recommended by Roger very nearly coincides with the
second one of Dieulafoy. He states that the proper loca-
.tion for operation is in the fifth interspace, about midway between the left nipple and the sternum, but a
little nearer the former than the latter landmark.‡ If the
heart is hypertrophied, and therefore situated at the lower
part of the sac, or if held there by adhesions which
prevent its being left in its usual position when the fluid
accumulation pushes the pericardium and diaphragm
down, it may be advisable to tap in the sixth intercostal
space. Of this physical condition of the heart the surgeon must endeavor to become cognizant before the operation is instituted. Sibson's vote would likewise be in
favor of the fifth interspace, for he says§ the trocar should
penetrate above the upper edge of the sixth cartilage,

---

* Dieulafoy on Pneumatic Aspiration, pp. 222–224.
† Op. cit., pp. 230, 231.
‡ Bull. de l'Acad. de Méd., p. 1279.
§ Reynolds's System of Medicine, vol. iv. p. 436.

more than an inch within the mammary line, and be directed slightly downwards so as to avoid the heart, which, in its healthy condition as to size, etc., has its lower border above the level of the fifth space. If the heart's beat is felt at a lower level than usual, he advises the insertion of the trocar in the epigastric region, between the ensiform cartilage and the seventh costal cartilage. This is justifiable, because the lower border of the fully distended sac is usually at about the level of the lower end of the xiphoid appendix; hence, tapping may be done at the middle of this part of the sternum, in the depression between it and the costal cartilages. Agnew considers* the sixth interspace, one inch to the left of the margin of the sternum, as the most accessible route. Braune says the safest position for puncture in order to avoid the pleura is in the upper angle, between the left edge of the sternum and the fifth cartilage,† which would be in the fifth interspace, close to the sternum. Adhesion of the pleura may not occur for a long time after the presence of large effusion; hence he thinks it well to keep near the sternum.

If we consider the points recommended, it is evident that they may be reduced in a general way to four different localities,—the fourth interspace, the fifth interspace, the sixth interspace, and the fossa between the xiphoid appendix and the cartilages of the false ribs as they ascend obliquely to form the lower part of the thoracic wall. After having determined which locality is best, the operator, if he choose either of the first three, must decide whether it is preferable to keep within the line of the mammary artery, or to go outward towards the nipple. The objection to puncturing near the sternum is the

---

* Agnew's Surgery, vol. i. p. 348.
† Topographical Anatomy, English edition, p. 106.

vicinity of the mammary artery; but, on the other hand, the probability of piercing the pleura, if a point farther to the left be chosen, renders the latter position to some extent objectionable.

I shall discuss these points in order, and endeavor to decide as to their relative merits. To make the subject intelligible, I have introduced the accompanying figure

Fig. 8.

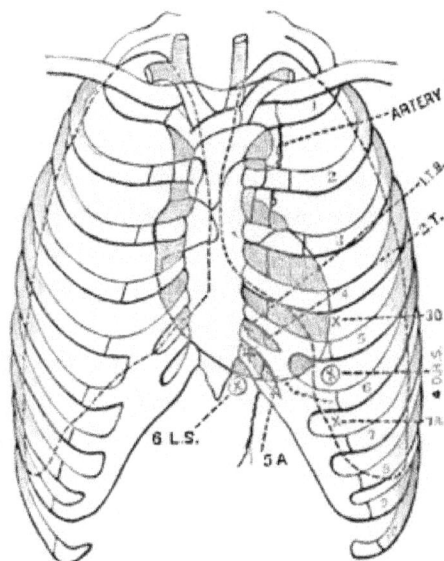

(Fig. 8), which shows the seven points mentioned by the authors quoted. Each point is indicated by a cross, to which is affixed the initial of the writer who recommends it; when he gives two points from which the operator is to choose, the initial letter appears in two places.

An examination of the figure shows a fact that might not suggest itself to one who is not in the habit of making accurate thoracic explorations. It is this, that the cartilages before joining the sternum curve upwards, and that

necessarily the intercostal spaces do the same; hence a
puncture made in one interspace near the sternum corresponds in level with a puncture made in the interspace
above, two or three inches to the left of the breast-bone.
Again, it must be recollected that the interspaces become
very narrow as they approach the median line, and that
the fifth and sixth, and sixth and seventh cartilages are
frequently joined by cartilaginous bands of varying width.
The seven points marked are in three instances close to
the sternum and within the ordinary line of the internal
mammary artery, in three instances far to the left of the
artery, and in one case between these two limits. The
reason for selecting the vicinity of the sternum is to avoid
puncturing the pleura, which is reflected upon the front of
the pericardium, but does not extend to the left edge of
the sternum.

Of the three internal points, I regard that between the
ensiform appendix and the seventh cartilage the best, because the two above are located in the narrowest portions
of the intercostal spaces, so that it is difficult at times to
push a trocar or needle between the cartilages. Even
if the instrument would pass it would be almost impossible to determine the proper point of skin at which to
introduce it, unless the preliminary incision, suggested by
Trousseau, was made, but which I reject, as mentioned
under the methods of operating. In a dried specimen,
before me as I write, there is only sufficient room in these
situations, and for an inch or more outwards, to allow a
large pin to be passed between the cartilages. There was
possibly more space before desiccation, but certainly not
sufficient to render this an available point for paracentesis.

An aspirating needle could, doubtless, in a young subject, be thrust through the cartilage, but a disk of cartilage would plug the instrument and prevent the fluid

flowing; then the operator would be likely to abandon the
operation or thrust the needle onward into the heart itself.
In the case of Ponroy (Case No. 33) the cartilage was
struck, though I do not know that he punctured in this
position, and a disk was found in the end of the needle on
its withdrawal; fortunately, there was room for the escape
of fluid by the lateral fenestra. Moreover, the pleura
approaches the edge of the breast-bone so closely that it
would very probably be injured, and if this occur the chief
advantage of puncturing near the sternum is lost. It is
interesting, however, to know that Baizeau after his experiments recommends the fifth space close to the sternum,
and believes the pleura would seldom be wounded.* There
is, however, plenty of space between the xiphoid appendix
and seventh costal cartilage, the fossa is quite easily felt
as a rule, and it is, moreover, lower down, and consequently nearer the bottom of the pericardial sac and farther from the ventricle than the other points.

If this point is selected, the trocar should be entered
close to the ensiform cartilage to avoid as far as possible
a small branch of the mammary artery, which traverses
this fossa along its internal surface. The pleura would
not be likely to be wounded, because it turns off to the
left before it extends down as far as the xiphoid fossa.
There is a possibility, however, of the puncture being
made too low down and the needle passing directly into
the abdominal cavity, for the diaphragm is attached to
the ensiform appendix and margin of the costal cartilages,
and then arches upwards, keeping for a time close to the
sternum. Hence, if the needle be not introduced as high
up in the xiphoid fossa as possible, it may pass below the
attachment of the diaphragmatic arch, or, if entering the

---

* Gaz. Hebdom. de Méd. et de Chir., 1868, p. 566.

pericardium, may by penetrating deeper project through the posterior wall, and go through the arch of the diaphragm into the abdomen and wound the liver. When the pericardium contains sufficient fluid to distend its walls and to depress the diaphragm, and when the patient is inspiring, there is of course less danger to be anticipated from these anatomical relations.

A locality between the three internal and the three external points is the one suggested by Agnew; but, though I hesitate to differ from so distinguished an anatomist and surgeon, I believe it to be bad. In the first place, it approaches too near the cartilaginous band, which usually joins the sixth and seventh cartilages, and, in the second place, in an instance where I tried it experimentally, the musculo-phrenic artery, one of the terminal branches of the internal mammary, and a vessel of considerable size, lay in dangerous proximity to the puncturing instrument. The distance between the puncture and the vessel was one-eighth of an inch. This may surprise the reader until it is recollected that the lower part of the breast-bone tapers and that the musculo-phrenic artery runs obliquely downwards and outwards; hence "one inch to the left of the margin of the sternum" is not outside the vessel. I believe that the pericardial cavity would be reached by such a puncture, but the objections mentioned serve to condemn it in my mind.

Let us next discuss the three external points delineated in the wood-cut. The one in the fourth interspace is too high up, while that in the sixth space is probably, for most cases, too low down, especially if it be carried farther to the left than the junction of the sixth rib with its cartilage. On these accounts I prefer that in the fifth intercostal space, which is situated very near the normal apex beat of the heart, close to and above the junction of the sixth rib with the corresponding cartilage. It is here, as well as in the fourth space, that the notch in the edge of the left lung occurs, which would prevent its being wounded even if the pericardial effusion had no tendency to push the lung away laterally.

The seven points given as proper for tapping the pericardium have now been reduced to two, viz.: the fossa between the ensiform and the costal cartilages of the left side, and the fifth interspace, near the junction of the sixth rib with its cartilage. Is there anything in favor of one of these over and above the other? A glance will show that the two positions are practically on the same level, but the diaphragm comes higher at the middle line than it does laterally; hence the former point of puncture would endanger this muscle more than the latter, though if high in the fossa there may not be much risk. The distended sac is said to be situated a little above, and at times even a little below, the point of the ensiform cartilage; hence the trocar thrust in no lower than the middle of this appendix would not be likely to endanger the diaphragm. The chief difference, it seems to me, resides in the fact that in one instance the pleural sac is not likely to be injured, while in the other case the trocar is pretty certain to perforate both layers of the left pleura before it enters the pericardial cavity, for there is not the notch in the pleura that there is in the border of the lung. In chronic pericarditis the pleural cavity in front of the pericardium will frequently be obliterated by inflammatory adhesions, but in hydropericardium the pleura will be normal, and some of the fluid may leak into the pleural sac during the operation or after the withdrawal of the needle. This will occur but rarely, and if it does, the absorbents of the healthy pleura will probably cause its disappearance in a short time, and with the aspirator there is no danger of pneumothorax occurring. If it were proposed to use a drainage-tube, it would be more important to be certain that the pleural cavity was obliterated at the point of operation, or else to select a point where the pleura was not likely to be wounded. Again, if we still used the old trocar without resorting to the suction power of the aspirator, this question would assume more importance; but aspiration empties the sac so effectually, and the needle makes such a comparatively small puncture in the pericardium, that leakage is not to be expected. That leakage may occur, however, is shown by Dr. Paul's case (No. 60). If there were a coexisting pleurisy, so that the pleural sac in front of the pericardium was also distended with serum, it might be judicious to attempt opening the pericardium at the inner situation, in order to avoid the confusion of tapping two superimposed cavities containing fluid. On this subject the reader is referred to the subsequent discussion of the complications of pericardial effusions.

In ordinary cases, then, I believe that the probability of puncturing the pleura is not a contra-indication to tapping the pericardium over the normal apex beat. The possibility, however, of wounding the diaphragm or the liver, which is prone to enlargement from venous engorgement in pericardial effusion, coupled with the fact that the ensiform cartilage is covered at its base by the overlying cartilages of the sixth and seventh ribs, making it difficult to determine its edge in fat subjects, renders paracentesis in the fossa between the xiphoid and seventh cartilage undesirable. Therefore I should tap in the former locality as a rule, reserving the latter for special cases where there was some indication for making an exception. Rotch's suggestion to tap on the right side of the sternum, in the fifth space, about four and one-half or five centimetres from the edge of the sternum, must be subjected to further clinical investigation before being accepted.

The method of operating would then be as follows: The patient should be as nearly recumbent as possible, in order to allow the heart to fall back from the anterior part of the pericardial sac. The intercostal spaces are then counted by recollecting that the first rib lies under the clavicle, or that the second rib joins the sternum at the prominent joint between the manubrium and gladiolus, and bearing in mind that the cartilages towards the median line approach each other and ascend obliquely. If the œdematous condition of the cellular tissue obscures the situation, hard rubbing of the chest may render the position of the ribs evident. The point is then selected in the fifth space, nearer the rib below than that above in order to avoid injuring the intercostal artery, and situated from two to two and a quarter inches (about five to six centimetres) to the left of the *median* line of the sternum, which, on account of the irregularity of the bone, is more readily determined than the border. It seems to me that this is far enough from the median line, and is better than a point nearer the nipple, because there is a possibility of making the puncture beyond the limits of the pericardium. In an instance where I made an experimental puncture on the cadaver in the fifth space, at a point two and one-half inches from the median line, the instrument entered the pericardial wall near its apex, and passed between the layers of the membrane without entering the cavity. Although this was in a case where there was no effusion, it shows that there is such a thing as going too far to the left; and this may especially occur in instances of solidification of the border of the lung complicating pericarditis, for then the diagnosis of the area of dulness, due to the pericardial effusion, is difficult to establish. It would be well to vary

the distance from the median line of the sternum to a slight
extent according to the patient's stature. In a child two
inches would be too far to the left; hence, the relation of
the median line, the apex beat, and the nipple must be
considered.

Care must be taken not to strike the costal cartilage lest
the point of the trocar be broken, or the needle plugged
with a disk cut from the cartilage. The caution is not in-
appropriate, because this has occurred, and is liable to take
place if the surgeon do not recollect that the cartilages have
a different direction from that of the corresponding ribs,
whose axis is *downwards* and forwards, but that of the cartilages *upwards* and forwards. As soon as the point of the
puncturing instrument is buried beneath the integument the
vacuum chamber of the aspirator should be attached, in
order that the glass index in the tubing may show by the
flow of fluid the instant the pericardium is opened. Unless
this is done, there is risk of pushing the needle into the
ventricular wall, and it is for this reason that an ordinary
trocar and canula should never be used, even if adapted to
the aspirating pump. If the skin is very thick, a prepara-
tory puncture may be made with a bistoury. It is better
in most instances to direct the needle backwards, but after
it has entered the sac its point may be turned a little downwards to avoid contact with the heart as it is thrown forward in systole. Should the dome-shaped trocar be used,
or that with the flexible end described by me, there is no
danger on this score. I should prefer to empty the sac
completely if possible, as this avoids the possibility of leak-
ing, though some may think it better to withdraw only a
portion of the effusion at first, and then repeat the operation. As the effusion decreases and the symptoms of dyspnœa ameliorate, which they generally do immediately, the
needle may be carefully retracted if there be fear of lacer-

ating the heart's surface. Finally, it is withdrawn entirely. The capillary puncture needs no treatment. If it
is thought necessary, a piece of adhesive plaster may be
placed over it. In cases where it is supposed that there is
floating lymph in the pericardium, the needle should be
larger than in cases of simple dropsy, when the finest
needle will allow withdrawal of the serum; hence in inflammatory effusions use an instrument of greater calibre.

I have thus discussed the method of performing paracentesis of the pericardium; it remains, then, to add some re-
capitulatory words. Be sure not to thrust the needle in too
deeply, for in emaciated subjects the thoracic wall is very
thin; and take care not to pierce a cartilage which may lie
underneath the point you have considered to be in an interspace. When a case shows evidence of pointing, as may
occur in chronic purulent pericarditis, the tapping should
be done at that point, without reference to its situation
in one or other of the localities discussed. So, also, if
the sounds of the heart be very loud at the point I have
advocated, and adhesion be inferred, the intelligent operator would select another point, where weakness of sounds
and intensity of dulness on percussion indicated that
the distance between the pericardium and the heart was
greatest.

It might happen that there was no flow of fluid after the
trocar was introduced through the integument, owing, perhaps, to the thickened pericardium being pushed in front
of the instrument; or, again, the flow might cease after
a few moments on account of the canula being choked by
a plug of lymph or inspissated pus. In the former instance,
it is safer to withdraw the needle and to ascertain that
it is not stopped by a disk cut from a cartilage, inadvert-
ently pierced, than to persist in thrusting it deeply inward.
Another puncture can then be made without the risk of

doing injury to the heart. In case of obstruction occurring from lymph, a wire may be introduced, or better still, if an aspirating trocar like that devised by me is used, the inner tube can be withdrawn, and the larger outside tube will probably be of sufficient calibre to allow the flakes of lymph to escape.

It is understood that a needle should never be used unless it is perfectly clean and not blocked up by rust. A case is reported* as follows: M. Potain having a patient with pericardial effusion, introduced the aspirating needle in the eighth space, but no fluid escaped though the instrument had penetrated to the depth of about three centimetres. The cardiac beat was felt against the instrument, and it was withdrawn, when it was found to be plugged with a piece of false membrane. No subsequent trouble supervened which could be attributed to the operation, but the patient died six days later. At the post-mortem examination it was discovered that the pericardium contained a litre of sero-purulent fluid. A different result might have occurred if the needle had been reintroduced.

In any event, great care should be exercised to prevent the admission of air; if this precaution be neglected a great advantage of aspiration is lost. I have frequently seen the aspirator used in such a bungling manner for thoracentesis, that the operator might as well have employed an ordinary trocar at the beginning.

---

* Le Progrès Médical, 1876, p. 76.

CHAPTER V.

DANGERS TO BE ENCOUNTERED.

THERE are but two dangers of the operation that need discussion at this point, since the question of wounding the pleura and diaphragm has been sufficiently considered previously. It is necessary to lay before the reader the possibility of hemorrhage from wounding the internal mammary artery, and to mention the danger of striking the ventricle and perhaps entering the cavity of the heart. The point of puncture that I have recommended precludes the possibility of injury to the artery, but as circumstances may induce the operator to select a situation near the median line, the course of the artery must be mentioned. After arising from the subclavian the internal mammary runs downwards, parallel with the edge of the sternum, crossing the inner surface of the costal cartilages, until it reaches the lower edge of the sixth cartilage; here, or in this neighborhood, it bifurcates into the superior epigastric, continuing directly downwards, and the musculophrenic, which runs downward and outward in the sixth interspace. The trunk in the region where it concerns us is somewhat less than three millimetres in diameter, and its two branches from say one and a half to two millimetres. The most important relation is its distance from the edge of the sternum, which fortunately is increased at the level of the fifth and sixth interspaces by the lessening width of the bone at its lower extremity. Cruveilhier and Sappey give as the average distance four to five milli-

metres,* Baizeau says it is ten to fifteen millimetres, while
Roger found in children a distance of only two to three millimetres. Some measurements of my own in adults show
it on the left side to be from one-quarter to one-half an
inch, which would correspond to about six to twelve millimetres, though in the neighborhood of the fifth and sixth
space I have seen it exceed this. Hence there is plenty of
room in adults to insert the trocar within the line of the
vessel, if it be kept close to the edge of the sternum. In
fat patients there may be difficulty in determining the edge
of the sternum with accuracy, and the artery may be accidentally wounded. This is the chief objection to the operation being performed at this locality. As stated previously,
the point suggested by Agnew seems improper, because I
found that it was dangerously near the musculo-phrenic
branch of the mammary. If this was found in one instance only, it renders the point more objectionable than
some others.

The proper method of avoiding the first danger, then, is
to keep away from the line of the vessel, and this is done
by puncturing in the fifth space about midway between
the sternum and the line of the nipple. If in any case
of operation by the aspirating needle the vessel were
wounded, it is probable that the surgeon would pass the
needle onward into the pericardium, and not be aware of
the division of the artery. There would be hemorrhage
into the areolar tissue of the anterior mediastinum, but it
is to my mind doubtful whether the amount would be very
great.

If detected, it should be treated by cold applications to
the chest, with ergot and astringents internally. Since
the method of preliminary incision and dissection has

---

* Roger, op. cit., p. 1271.

been abandoned, there is little risk of hemorrhage from the mammary artery, if a proper amount of care be taken in selecting the point of puncture.

An accident of much more importance is injury done to the heart, which may occur from thrusting the trocar into the substance, or even into the cavity of the heart. A mere abrasion or scratch of the surface of the organ is of little moment, for its only tendency would be to cause a slight local pericarditis, which would not aggravate the existing condition. It may happen, however, that the operator's instrument pass into the tissue of the ventricle before he is aware of it, and that he push it onward, in his search for fluid, until pure blood escapes from the cardiac cavity. That this is not an impossible accident is shown by Roger, who believes that his case, numbered 29 in the table, may have shown blood at first, because the ventricle was punctured. The same question also arises in considering the case of Baizeau, where clots were found in the pericardial cavity at the autopsy. It must be admitted, however, that no wound of the heart was visible in either instance when the autopsy was made, though in the latter case the patient died a couple of hours after the operation.

Roger's third case is related by him as a genuine instance of wound of the heart. I shall, therefore, give it more in detail than in the tabulated account. The boy was five years old, and suffered from pericarditis. When the aspirator was introduced there was at first no fluid found; then there escaped blood mixed with serosity; then pure blood resembling venous blood, and this flowed in a steady stream without pulsation. Two hundred grammes were thus evacuated when the child became pale, and the needle was withdrawn. Subsequently, it was determined that there was less prominence of the chest and less dulness, and the cardiac sounds were more superficial; but the child was pale, was sweating, and had an imperceptible pulse. Improvement and cure followed, and the pericardial symptoms did not return, though the case was under observation for three months. The antecedent organic trouble of the heart itself, however, progressed, and death took place about five months after the tapping. The autopsy showed dilatation of the cavities of the heart with mitral insufficiency, and general pericardial adhesion. Roger himself says that the blood came directly from the right ventricle.

In this connection a most remarkable case, reported to the Clinical Society of London,* may be mentioned. It is the same case related in a preceding chapter. A woman, aged twenty-seven years, had pleuro-pneumonia and signs of large pericardial effusion; as she was almost moribund a trocar was introduced by Mr. Hulke through the fourth interspace, but, to the dismay of the surgeon, dark, venous blood escaped. The instrument was immediately withdrawn, but the patient, instead of showing unfavorable symptoms, seemed to be relieved of the distress and dyspnœa. She died four weeks later of a complication of diseases, when the autopsy showed dilatation and valvular disease of the heart, with the pericardium universally adherent, but no effusion. These two instances prove that a wound of the heart is not impossible even in capable hands, and also tend to show us that there is not as much danger of fatal syncope occurring under these circumstances as we should suppose. In fact, in the latter case, although the right ventricle was tapped and one drachm of blood withdrawn, the patient exhibited no shock or distress; but the abstraction of blood

---

* Transactions Clinical Society, viii. p. 169.

seemed to relieve the distended heart much better than phlebotomy would have done, as was evinced by the diminution of threatening symptoms and the decrease of the area of dulness.

Though no great fatality seems to attend this accident, it is one to be avoided with scrupulous care; hence the propriety of selecting a low point of puncture near the apex of the heart, rather than towards the base, where the space between heart and wall of thorax is less. Another objection to a high puncture is the much graver prognosis attending wounds of the thin-walled auricle. Dr. Steiner in his experiments with electro-puncture found that needles could be quite safely introduced into either ventricle, provided they were at once withdrawn;* but it is not so safe to puncture the auricles. Other observers, such as Cloquet, Bouchut, Legros, and Onimus, have noticed the apparent innocuousness of wounds of the heart made by capillary trocars. A case bearing upon this point is recorded, in which a needle had accidentally entered the chest and penetrated the heart to some depth. With the exception of pain no symptoms of note were observed, and the foreign body was shortly afterwards removed without causing the slightest local or constitutional disturbance. Again, a needle has been found firmly fixed across the ventricular septum, which had evidently been there a long time, for it was coated with lymph, and death had been the result of other causes.†

In this connection I must mention an instance where death occurred apparently from the operation of paracentesis, though the aspirating needle did not wound the heart

---

* Med. Times and Gazette, May, 1873, p. 492; from Langenbeck's Archiv für Klin. Chirurgie.

† Med. Times and Gazette, May, 1873, p. 491.

itself, and failed to reach the effusion. A man was supposed to have effusion in the pericardium, and aspiration was attempted in the third interspace, but no fluid escaped, and he died almost immediately. The autopsy revealed the fact that the instrument had pierced not the heart wall as might be expected, but had penetrated the thickened pericardium at a point where it was adherent to the heart. The pericardium was adherent to almost the whole of the left heart, while at the right there was a cavity containing twelve hundred grammes of fluid. There was also valvular disease.* How to explain this sudden death from shock I do not know. The operation was done higher up than usual, and death may have been due to some interference with cardiac action resulting from injury to the auricle with its adhering pericardium, that had assumed the functions of the auricular wall. It is possible that the man, enfeebled by valvular disease and the pressure from the large effusion, was unable to withstand the mental and physical shock produced by the operation. I have omitted this case from the table of cases of paracentesis pericardii because the pericardial cavity was not opened, and therefore it could not be classed among them. The importance of the case requires its insertion here in order that I may present a scientific and non-partisan view of the subject in all its bearings.

The experiments of Steiner and others have been quoted, not to encourage careless employment of surgical means in treating pericardial effusions, but to let it be known that we should throw aside the foolish prejudice that makes us shun operative interference in pericarditis because of the supposed vulnerability of the heart. If the cases of peri-

---

* Bull. Gén. de Thérap., 1878, tome xciv. p. 428; from Union Méd. and Gaz. des Hôp., 1878, p. 310.

cardial effusion be properly selected, and diligent care be exercised in the performance of the operation, the risk of striking the heart is very slight. If the heart should be punctured through some unforeseen contingency, such as an unsuspected adhesion, it will probably, if we are to judge by the cases quoted, add little to the gravity of the case, which, to say the least, is already very grave when the operation is suggested. Hence I conclude that the fear of wounding the central organ of circulation should not deter the surgeon from tapping the pericardium, if the diagnostic signs warrant the attempt. He must be careful that no disk of cartilage or shred of membrane occludes his needle, and then he can push it onward until the index shows the fluid escaping. In thick, œdematous, or fatty walls it may be necessary to go to a depth of four or five centimetres before reaching the fluid; hence it is absolutely essential that the needle's patulency be assured.

OBJECTIONS TO THE OPERATION.

I deem it proper to devote some time to considering the objections that have been advanced against the adoption of paracentesis pericardii as a justifiable procedure. In the first place, it has been said almost with a sneer that the operation is merely palliative. Now, even if I were willing to admit this, which I am not, I should still insist upon the performance of paracentesis of the pericardium in proper cases. In many incurable affections we must labor with all our might to palliate the sufferings of the patient. If it were right to do so, the surgeon's feelings of humanity might prompt him to hasten dissolution when he sees and hears the agony of a man with hydrophobia; but no, he must palliate. He can destroy a brute to relieve prolonged suffering, but he must sustain the life of a human being. Hence palliative measures are urgently demanded in every department of medicine. If a few days can be added to a life by tapping the pericardium, it should be done; aye, and done quickly. Allbutt, when discussing the objection of one who says, " In the majority of cases I believe the result has been unfavorable,"* argues, very properly, that " unfavorable" must mean that the operation itself caused death, hastened the fatal issue, or augmented the suffering of the patient while doing no good whatsoever.† An examination of the accompanying table shows that these effects are not chargeable to the operation, but that it, as a rule, relieves suffering at once and prolongs life.

This feeling of hostility to the operation has of late years been considerably modified, and I doubt whether Billroth would now say that the operation seemed almost like prostitution of surgical skill, or speak of its resemblance to a surgical frivolity.‡

Jaccoud speaks of the operation as legitimate at times, but says that, after the performance of paracentesis of the pericardium, the liquid is reproduced with great facility on account of the diminution of extra-vascular pressure.§ This statement I am unwilling to accept as a clinical fact, and, even if true, I should not feel that it carried much weight as an objection to the operation.

As an answer to those who disparage the operation because of its merely palliative character in many instances, I will state that in the table of 60 cases of operation there are recorded 36 deaths. Of these fatal cases, 23 patients are known to have survived the operation for one day or

---

* British Medical Journal, July 2, 1870.

† Ibid., July 9, 1870, p. 32.

‡ Handbuch der Allgemein. und Speciellen Chirurgie, III. Bd., II. Abt., I. Lf., 163 S.

§ Pathologie Interne, i. 646.

more, 9 are known to have lived less than a day after the operation, and in 4 instances the time of survival is not given. The average time of survival of those who are known to have lived one day or more is over 27 days. The greatest length of time was 160 days. If it be recollected that most cases are not considered proper for operation until the patient is almost moribund, this is certainly a good showing. Four weeks is a long time to add to a man's life, especially when it can be done by an operation that causes so little pain.

The probability of adhesion of the two layers of the pericardium occurring after tapping has been held as a contra-indication, because adhesions may induce, it is said, valvular disease or pathological changes in the cavities of the heart. If such is the ultimate effect of adhesion, it is better than the existing condition of pericardial effusion, which gives rise to myocarditis and pathological sequences of more immediate danger. That general adhesion does take place in inflammatory effusions is probable. Kyber found that it had occurred in three of his cases of recovery after paracentesis, where autopsies were obtained long afterwards ;* and Pepper's case (No. 57) showed the same condition when the post-mortem examination was made fifteen months after the operation. It would seem that a radical cure is to be looked for in the occurrence of union of the two surfaces, and if this be true, the objection to the operation on the score of pericardial adhesion falls. After all, the degree of agency exercised by adhesions in the production of other cardiac lesions is a disputed matter, as will be seen by careful perusal of the works on heart affections. Are we to reject an operation which prevents imminent death because we fear it will lead to a condition

---

* Monthly Retrospect of Medical Science, Edinburgh, March, 1848.

that has a doubtful agency in producing other less fatal ailments?

An objection of more force than either of those thus far reviewed is the assertion that the fluid accumulates with greater rapidity after tapping than previously, and that it has a tendency to become purulent. We have not sufficient data to answer this point, but I do not see that the objection is of any more value than against aspiration in pleuritis, where, if the fluid reaccumulate, the needle is introduced again and again, and cure finally effected; and if the pleural effusion become purulent, it is evacuated either by repeated tapping or by continuous drainage. The proper method of obviating suppuration is to avoid the entrance of air into the cavity during the operation. The line of treatment which meets this indication, and the course to pursue after pus has formed, will be found in the chapter which treats of the methods of operating. In Frerich's ward, where thoracentesis is frequently performed, no serous effusion, it is stated, becomes purulent if the instrument be disinfected and the air excluded from the pleural cavity.* This would probably correspond with general experience. Is there any difference in this regard between pleura and pericardium?

There can be no valid objection urged against a second operation if the first has been deemed justifiable. A study of the table of operations will furnish information concerning this matter. The cases of Schuh and one or two others must be omitted, since in them the first tapping failed to give exit to the fluid, and a second operation was immediately performed.

It is stated that the operation was done more than once, as follows:

---

* Medical and Surgical Reporter, September 30, 1876, p. 274.

There were tapped twice	.	.	.	.	.	.	11 patients.
" was " three times		.	.	.	.	1 patient.
" " " six times	.	.	.	.	.	1 "
" " " eight times		.	.	.	.	1 "

There was a repetition of the operation therefore in   14 patients.

The patient who was tapped six times (No. 43) lived thirty-eight days after the first operation, or ten days after the sixth, and finally died, having peritonitis in addition to pericarditis. Bouchut's patient (No. 48) was tapped eight times, and lived thirty-four days after the primary operation. Death occurred three days after the final tapping, when the heart itself was punctured. The shortest interval between the original operation and the second was rather less than one day; in other cases the period is shown by the full histories to have been as long as fourteen, fifteen, and seventeen days.

Let us return, however, to the questions: Does paracentesis itself cause rapid reaccumulation? and, if so, Is the second operation more dangerous than the first? In the first place, there are twenty-three cases reported where recovery followed paracentesis without a second operation being necessitated, and in the fourteen instances where it was required there was additional disease in every case. Secondly, of the fourteen cases of repeated tapping thirteen died; but in all of them there was either disease of the heart or lungs, as in nine, scurvy or abdominal disease, as in three, or brain disease, as in one; and, indeed, the one patient who recovered after double tapping had phthisis (No. 19). These statistics seem to show pretty conclusively that repeated tapping is not demanded as a sequel of first paracenteses, but is required because the patient's condition causes a spontaneous reaccumulation, which would occur if the effusion was suddenly removed by any other method that did not at the same time improve his

diathesis.   They also militate against the idea that there
is decided risk in tapping more than once, for, though it
is not proved that these patients died of the accompanying
disease, yet it is shown that the fatal cases, where repeated
paracentesis was performed, were decidedly unfavorable.
Therefore the evidence is of value, though it be negative.
More positive evidence is the fact that, in nine of the thirteen
fatal cases of repeated tapping, the time of survival after the
second operation was one day or more, while in some cases
the patient survived many days after the second operation.
In the four remaining instances death occurred in less than
a day, or the time is unknown.   There is no proof in this
that a second operation is to be dreaded any more than the
primary one, except in so far as it suggests the probability
of some affection complicating the pericardial effusion.

TREATMENT OF COMPLICATIONS.

Having carefully taken up in detail the steps to be pursued in treating pericardial effusions, I shall occupy a few
pages with some consideration of the management of complications that may exist as causes or results of the effusion
in the pericardial cavity.   Pleuritis with effusion is a
not uncommon accompaniment of pericarditis, as has been
shown when discussing the ætiology of inflammation of
the heart's investment; so, also, pneumonia is frequently
found in a similar relationship.   These affections must be
treated on the general principles that guide the physician
in managing cases unaccompanied by pericardial disease.
The difficulty of diagnosis that may arise between left
pleuritic effusion, especially if encysted, and effusion in
the pericardium has been mentioned.   Occasionally it is
hard to calculate the degree of causation exerted by these
two effusions in producing the symptomatic manifesta-
tions observed.   There exist certain symptoms, such as

cyanosis, dyspnœa, feeble circulation, etc.; are they due to the serum in the pleura or to that in the pericardium? When the pleural effusion is relatively much greater in amount, the question can be decided; but cases arise in which it is almost impossible to say which is the more active agent. If there be any doubt, it is proper to perform thoracentesis first, and then await results; if relief does not follow, the pericardium is to be punctured subsequently. There is no objection to doing both operations at one time, if they are indicated by the condition of the patient.

The treatment of renal symptoms must be governed by the general indications in much the same manner; but I wish to enforce attention to the fact that albuminuria and uræmic convulsions may be due to the pressure from the pericardial effusion, and that the surgeon should not hesitate to tap because he finds evidences of kidney trouble. Often there is hydro-pericardium as a part of a general dropsical condition, induced by renal disease, but the reverse may be the proper relationship. In other words, what is cause in one instance may be effect in another. The nephritic congestion or inflammation is to be treated by diuretics, such as digitalis, scoparius, juniper, and by derivatives, such as jaborandi; but it is useless to mention measures or methods of medication, since this belongs to general clinical medicine, and not to the subject of our present volume.

There have been a few cases recorded where a pericardial fistula has been established either by a spontaneous or operative evacuation of the effusion. Some cases so called have doubtless been pleural fistules, or superficial sinuses due to a subtegumentary abscess. If they are truly pericardial, I advise dilatation of the orifice with compressed sponge, and washing of the pericardial cavity

with astringent and disinfectant solutions. It may be that
the layers of pericardium are adherent, and that the sinus
opens into a small suppurating pocket between the visceral
and parietal membrane; or this pocket may be external to
the pericardium entirely, and be formed in the midst of
some new tissue occupying the mediastinum. It is, of
course, useless to expect closure of the external opening
if there is a continual secretion of pus within, and it
may be necessary to lay open such cavities. Sinuses
that are superficial demand similar treatment. They
should be laid open with the bistoury, and forced to gran-
ulate from the bottom. It is possible that such pathological conditions may depend for their existence upon a
diseased rib or cartilage; if such be the fact, resection, or
removal of the diseased tissue with the burr or chisel, is
to be considered.

CHAPTER VI.

TABLE OF CASES.

AFTER an extended search through very many volumes
of text-books, monographs, and journals, and after a good
deal of correspondence, I have collected sixty cases of paracentesis of the pericardium, and have embodied them in
the following table. Some cases that I have found mentioned, or have heard of, I have rejected, because I could
not get any definite information of them, or because I believed them to be of doubtful authenticity. I have person-
ally examined many files of German, French, Italian, and
Spanish journals in order to get trace of the cases operated

upon throughout the continent of Europe; and the fact
that I have again and again met references to cases already tabulated, seems to me conclusive evidence that my
search was pursued with a considerable degree of thoroughness, and that I have missed very few recorded operations. It may be, and probably is, a fact, that I have
failed to collect all; but a prolonged study of indexes and
a watchful eye on current literature during a number of
years have probably given me an opportunity of finding
nearly all the published operations. It may be observed
that some of the cases, that appeared in a table published
by me a few years ago, are changed in certain respects
in the present work. This is due to my having gained
more reliable information respecting the points of interest. In this table I have omitted Bowditch's case, which
was placed in the other on the authority of Trousseau.
Dr. Pepper states that Bowditch says he never performed
the operation. When compiling the first series of cases I
wrote to Dr. Bowditch, of Boston, but never obtained any
answer from him, owing doubtless to my letter being mis-
laid, for I know that he was away at the time.

Some of the cases have two names attached. This is
due to the fact that occasionally a case is reported in
different journals under a different name, owing to the
journal which makes the abstract or excerpt getting the
names of the operator and attending physician confused.
In this way I have several times met the same case under
a different heading, and have thought I had found a new
instance of the operation, until comparison of the dates,
ages, results, etc., has proved that I was in error. To
avoid misapprehension I have added the second name in
parenthesis.

Operator.	Date.	Sex and Age.	Mode and Site of Operation.	Recovery.	Death.	Time that the Patient survived the Operation.	Remarks.	Complication.	Reference.
1, Romero.	Before 1819	M. 35	Bistoury and scissors. 5th interspace.	1					Dict. des Sciences Médicales, Paris, 1819, xl. 371.
2, Romero.	Do.	M. 37	Do.	1					Do.
3, Romero.	Do.	M. 45	Do.						Do.
4, Jowett.	1827	F. 14	Not stated.	1	?	Life prolonged.	Hope of recovery.		Günther, Blutigen Operationen, iv. 3, 102.
5, Kurawagen.	1839	M.	Trocar. 5th interspace.	1			Hemorrhagic scorbutic pericarditis. Drew off 60ijss. Quite well five months later. Scorbutic pericarditis.	Scurvy.	British and Foreign Medical Review, July, 1>41.
6, Karnwagen.	1839	M.	Do.?		1	Life was prolonged. 100 days.		Scurvy.	Do.
7, Schuh.	1840	F. 24	Trocar. 4th interspace.		1		Tapped first in third interspace. Case was one of encephaloid disease of thoracic viscera.	Cancer.	Archives Générales de Médecine, November, 1854.
8, Kyber.	1840	M.	Trocar.* 4th interspace.	1			Scorbutic pericarditis.	Scurvy.	Monthly Retrospect of Medical Sciences, Edinburgh, March, 1848, i. 35.
9, Heger.	1841	M. 19	Trocar. 5th interspace.		1	60 days.	Tapped twice. 1500 grammes and 400 grammes. Drainage-tube left in six hours.	Phthisis.	Archives Générales de Médecine, November, 1854.
10, Schönberg.	1842	M.	Trocar.	1			Hemorrhagic effusion. Removed 5 lbs. Recovered in six weeks.	Scurvy?	Günther, Blutigen Operationen, iv. 3, 102.
11, Kyber.	1843	M.	Trocar. 4th interspace.	1			Scorbutic pericarditis. Was living one and a half years later.	Scurvy.	Do. and also Monthly Retrospect of Medical Sciences, March, 1848, i. p. 35.
12, Kyber.	1845	M.	Do.	1	1	17 days.	Scorbutic pericarditis. Tapped twice, with interval of 17 days. Scorbutic pericarditis.	Scurvy.	Do.
13, Kyber.	1845	M.	Do.						Do.

* Sometimes Kyber adapted a syringe to the trocar.

Operator.	Date.	Sex and Age.	Mode and Site of Operation.	Recovery.	Death.	Time that Patient survived the Operation.	Remarks.	Complication.	Reference.
14, Kyber.	184-	M.	Trocar. 4th interspace.	1			Scorbutic pericarditis.	Scurvy.	Günther, and also Monthly Retrospect of Medical Sciences, March, 1848, i. p.35.
15, J. C. Warren.	1852	F. 35	Incision and trocar. 6th interspace.	1			Removed f. oz. v. Left hospital in a few weeks.		H. H. Smith's Surgery, ii. 358.
16, Champenillon.	1849	M.	Trocar. 4th interspace.	1			615 grammes, 18 months after was working as a sailor, Roger says this was a complete cure.		Bull. de l'Académie de Médecine, 1875, p. 1266.
17, Jobert.	1854	M. 16	Incision and trocar. 5th interspace.	1			Removed 400 grammes. Tapped pleura also for effusion. Under notice three months.	Phthisis.	Trousseau, Clinical Medicine, iii. 350.
18, Béhier.	1854	F. 22	Trocar. 6th interspace.		1	26 days.	Removed 250 grammes. Tapped previously in 7th intercostal space; no fluid obtained.	Diet of pneumonia.	Archives Générales de Médecine, November, 1854.
19, Aran.	1855	M. 23	Incision and trocar. 5th interspace.	1			Tapped twice. F. oz. xlix. Injected f. oz. xlix and f. oz. iodine and iodide of potassium after each tapping.	Phthisis.	Trousseau, Clinical Medicine, iii. 386.
20, Aran. 21, Aran. 22, Stœhr.			Do. Not stated. Not stated.	1 1	1				Id. iii. 391. Id. iii. 391. Id. iii. 383.
23, Vernay.	1855	M. 23	Trocar. 5th interspace.		1	21 days.	Tapped twice. First, 500 grammes; second, three days later, 400 grammes. Tapped abdomen for ascites.	Valvular disease.	Half-Yearly Abstract of the Medical Sciences, xxv. p.95.
24, Trousseau.	1856	M. 27	Incision.		1	5 days.	Removed f. oz. iij? Tapped pleura accidentally at same time.	Pleurisy and phthisis.	Trousseau, Clinical Medicine, iii. 364.
25, Wilezkowski.	1857		Incision.		1	6 hours.	Hemorrhagic pericarditis.	Scurvy?	Günther, Blottgen Operationen, iv. 3. 102.

Operator.	Date.	Sex and Age.	Mode and Site of Operation.	Recovery.	Death.	Time that Patient survived the Operation.	Remarks.	Complication.	Reference.
26, Wheelhouse.	1866	M. 26	Trocar, 4th interspace.	1			Removed f. oz. ij. Living 23 months later. Acute rheumatic pericarditis.	Rheumatism.	British Medical Journal, Oct. 10, 1868.
27, Roger.	1869	F. 12	Trocar, 6th interspace.		1	1 day.	Removed 780 grammes. At first no flow. Introduced again lower down.	Myocarditis and heart clot.	Bull. de l'Académie de Médecine, 1875, p. 1284.
28, Mauler (Loebel).	1868?	F. 68	Aspiration. 3d interspace.		1	15 days.	Tapped twice. First, f. oz. ij. Second time at night of sternum.	Pleurisy.	Wochenblatt der K. K. Geschschaft der Aerzte in Wien, 1868, No. 24, 221 S.
29, Roger.	1868	F. 11	Trocar, 5th interspace.		1	35 days.	Tapped twice. First, 100 grammes (blood); second time, 500 grammes serum, five days after.	Ascites.	Bull. de l'Académie de Médecine, 1875, p. 1274.
30, Teale.	1869	F. 27	Trocar, 4th interspace.		1	Few hours.	Tapped twice. Removed f. oz. v and f. oz. vj two days later.	Phthisis?	Lancet, June 12, 1869.
31, Baizeau.		M.	Incision, 6th interspace.		1	2 hours.	400 grammes blood. Found at autopsy lymph covering heart and pericardium. Sanguinolent serum and clots. In pleura 450 grammes, apparently from pericardial cavity. Wound of pericardium seen, but no wound of heart found. Resembles a case of wound of heart, but Baizeau thought it was not, because no wound was found, and the fluid did not escape in jets.	Operation at a previous time for sub-maxillary adenitis.	Bull. de l'Académie de Médecine, 1875, p. 1273; Gaz. Hebd. de Médecine et de Chirurg., 1868.
32, Herbert Norris.	1869	F. 14	Bowditch's pump and cannula. (Aspiration) 4th interspace.		1	22 hours.	f. oz. ijss serum.	Acute rheumatism, acute endocarditis, and pleuritis.	Agnew's Surgery, i. p. 348. Letter from operator.

Operator.	Date.	Sex and Age.	Mode and Site of Operation.	Recovery.	Death.	Time that Patient survived the Operation.	Remarks.	Complication.	Reference.
33, Pouroy (Frémy).	1870	M. 21	Aspiration. Too much œdema to count ribs.	1			800 grammes sero-pus. Passed from notice after 50 days, but had symptoms of chronic lung disease. On withdrawal of needle found piece of cartilage in the end.	Left pleuritis.	Bull. Gén. de Thérap., 1871, tome lxxx. p. 125; Dieulafoy on Pneumatic Aspiration, London, 1873, p. 215.
34, Labric.		M. 6	Trocar. 5th interspace, four centimetres outside nipple.	1			Intended to tap left pleura, as diagnosis was pleuritis and pericarditis. F. oz. xvj purulent serum flowed, but autopsy showed it had come from pericardium, and that pleura was adherent. Reporter says that death must not be attributed to the tapping.		Bull. de l'Académie de Médecine (1873), t. xl. p. 1216; also, id., t. xxxvii. p. 658.
35, Roger.	1872	M. 5	Aspiration. 6th interspace.		1		At first no fluid, then blood and serum, followed by pure blood, apparently venous, and not in jets. 200 grammes. Child became pale. Improvement followed. Pericarditis did not return. Death occurred five months after operation. Author says the right ventricle was wounded.	Dilatation and valvular disease of heart.	Bull. de l'Académie de Médecine, 1873, p. 1276.
36, Chaillou.		M.	Aspiration.		1	14 days.	First tapped left pleura, then aspirated pericardium, and withdrew 625 grammes serum. Ten days later aspirated pleura again, and injected iodine. Sputa like phthisis.	Left pleuritis.	Dieulafoy on Pneumatic Aspiration, London, 1873, p. 241.

7

Operator.	Date.	Sex and Age.	Mode and Site of Operation.	Recovery.	Death.	Time that Patient survived the Operation.	Remarks.	Complication.	Reference.
37, Juergensen.	1872	M. 6	Thiersch's syringe (Aspiration) at first. Trocar second operation. Site not given.		1	7 days.	First tapping, 3 ounces pus. Second operation, two days after, 6½ ounces pus. Left cannula in sac, and washed out with salt solution. Died with cerebral symptoms. Followed pneumonia of left side.	Cerebral meningitis.	Ziemssen's Cyclopaedia of Medicine, Am. ed., vol. v. p. 113.
38, Duncan.		M. boy	Trocar.		1	Few hours.			
39, Chairou.	1872	M. 23	Aspiration. 5th interspace.		1	49 days.	1000 grammes serum. Tapped pleura 1430 grammes. Died in forty-nine days of diarrhœa and phthisis.	Diarrhœa and phthisis.	Edinburgh Medical Journal, Oct. 1872, p. 376. Bulletin de l'Académie de Médecine, 1872, tome xxxvii. p. 1011.
40, Maclaren.	1872	M. 27	Incision and trocar. 5th interspace.		1	6 days.	Removed f. oz. xxxv.	Pleurisy.	Edinburgh Medical Journal, June, 1872.
41, Heath.	1873	M. 6	Aspiration. 3d and 4th interspaces.		1	50 days.	Tapped pericardium twice. f. oz. iij¾ and f. oz. vj. last time in 4th interspace. Tapped abdomen twice.	Phthisis and tubercular peritonitis.	Practitioner, xi. 265.
42, Saundby.	1874	M. 13	Aspiration. 4th interspace.		1	Few hours.	Removed pus f. oz. xxx; probably from rupture of pulmonary abscess.	Pleurisy and abscess of lung.	Edinburgh Medical Journal, March, 1875, p. 799.
43, Gooch.	1874	M. 13	Aspiration. 5th interspace.		1	38 days.	Tapped six times. Purulent fluid. F. oz. xxxj; f. oz. xxxv; f. oz. ix—iodine injected; f. oz. l—iodine; f. oz. xxx; f. oz. xx (?)—iodine.	Peritonitis.	British Medical Journal, June 19, 1875.
44, Steele.	1874	M. 25	Aspiration. 4th interspace.	1			Acute rheumatic pericarditis.		Lancet, Aug. 22, 1874, p. 271.
45, Bartleet.	1874	M. 20	Aspiration. 4th interspace.	1			Acute rheumatic pericarditis. Removed f. oz. xiv. Walking about in twenty-seven days.		Lancet, December 19, 1874.

Operator.	Date.	Sex and Age.	Mode and Site of Operation.	Recovery.	Death.	Time that Patient survived the Operation.	Remarks.	Complication.	Reference.
46, Nixon.	1876	M. 20	Aspiration. 5th interspace.		1	6 days.	Removed f. oz. iijss.	Pleurisy.	Dublin Journal of Medical Sciences, June 1, 1876.
47, Villeneuve.	1873	M. 5	Aspiration. Most projecting point.	1			Two syringefuls of serum. Puncture did not close, fluid became purulent, and flowed for nearly six months. Finally closed spontaneously.		London Med. Record, Sept. 15, 1875, p. 572; from Marseille Médical.
48, Bouchut.	1873	F. 11½	Aspiration. 5th interspace.		1	34 days.	Tapped eight times at same place. Serous effusion becoming hæmorrhagic. Two punctures of the heart without accidents. Death followed 3 days after eighth operation. Intervals between the operations were 4 days, 3 days, 3 days, 4 days, 7 days, 7 days, 3 days. Autopsy showed nearly 900 grammes of chocolate-colored fluid in pericardium. No syncope or aggravation seemed to follow the punctures of the heart, though 80 grammes and 30 grammes of blood were withdrawn.	Left pleuritis 6 weeks previously; endocarditis.	Gaz. des Hôpitaux, 1873, p. 1130.
49, Macleod.	1874	M. 23	Aspiration. 5th interspace.		1	29 days.	Tapped three times. Oz. xx serum; 15 days later, oz. xxx; 11 days subsequently, oz. xv.	Pneumonia and pleurisy.	Glasgow Medical Journal, July, 1877, p. 361.
50, Lyon.	1874	M. 31	Trocar. 5th interspace.		1	9 days.	Tapped twice at interval of 7 days. Oz. xxxii and oz. iss pus.	Pneumonia.	New York Med. Record, April 1, 1876, p. 221; from Med. Comm. of Connecticut Med. Soc., 1875.

Operator.	Date.	Sex and Age.	Mode and Site of Operation.	Recovery.	Death.	Time that Patient survived the Operation.	Remarks.	Complication.	Reference.
51, Valtosta.	1874	M. 35	Incision and trocar. 5th interspace.		1	28 days.	Paracentesis of chest twice, and abdomen once, previously. Oz. x fluid from pericardium. Subsequently pleura tapped again.	Pleuritis; ascites.	London Med. Record, iii. p. 275; from Giornale Veneto di Scienze Mediche, 1875.
52, Welch.	1875	M. 45	Aspiration. 4th interspace.		1	5 days.	Oz. xxviii. pus. Great relief. Respirations fell from 60 to 28. Patient was so much exhausted the operator did not operate a second time, though he considered the question.	Pleuritis when attack began?	Trans. State Med. Society of Arkansas, 1875–76; quoted in American Journal of Med. Sciences, January, 1877, p. 190.
53, Elliott (Burder).	1875	M. 60	Aspiration. 5th interspace.	1			Oz. xliii serum. Left hospital in ten weeks greatly improved. Several months later was in hospital again with cardiac symptoms, but there was no indication of fluid in the pericardium.	Chronic cardiac disease.	Lancet, January 8, 1875, p. 50; Trans. Bristol (England) Med. Chirurg. Society, vol. i, 1878, p. 75.
54, J. Lewis Smith ?	1875?	— 13	Aspiration.		1		Oz. xxx pus. After death found oz. xxxvi in pericardium. Pericardial effusion probably developed by admission of pleuritic effusion through a minute opening.	Pleurisy.	New York Medical Record, February 12, 1876, p. 110.
55, Viry.	1876	M. 22	Aspiration. 4th interspace; 2d operation trocar, 4th interspace.		1	22 hours.	Tapped twice. Pus 43 grammes. Next day used large trocar and pump because fluid so thick. Pus and clots 140 grammes. Left canula in place, intending to inject iodine, but did not do so.	Right pneumonia.	l'Union Médicale, 25 Fév, 1879, p. 315.

Operator.	Date.	Sex and Age.	Mode and Site of Operation.	Recovery.	Death.	Time that the Patient survived the Operation.	Remarks.	Complication.	Reference.
56, Hunt (?)	1877	F. 16	Aspiration. 4th interspace?		1	2 days?	Few drachms. At autopsy pericardium contained oz. xiv. Near juncture firm adhesion to heart, which may account for small amount of fluid obtained.	Rheumatic fever.	Lancet, March 10, 1877, p. 341.
57, Pepper.	1877	F. 17	Aspiration. 5th interspace.	1			F. oz. viij serum. Able to get out of bed 26 days after operation. Pleuritis and ascites 3½ months subsequently. Death occurred 15 months after operation. Autopsy showed complete adhesion, no valvular lesion, fatty degeneration of muscular structure.	Albuminuria and casts; due to effusion in pericardium.	Medical News and Library, March, 1878, and American Journal of Med. Sciences, April, 1879.
58, Douglas.	1878	M. 46	Aspiration. 4th interspace.	1			F. oz. liv pus. Living when case reported 30 days subsequently. Expectorates purulent matter, though his condition is far better than at operation.	Pulmonary disease?	Trans. Detroit Medical and Library Association, January, 1879.
59, Porcher.	1878	M. 50	Hypodermic syringe. (Aspiration.) 4th or 5th interspace.	1			F. oz.] sero-sanguinolent fluid. Much relief. Subsequently tapped right pleural cavity twice. Died several months after with general dropsy. Oj fluid in pericardium.	Pleuritis and anasarca.	New York Medical Record, July, 1878, p. 45. Letter from operator.
60, Comegys Paul.	1879	M. 22	Hypodermic needle. 4th interspace.		1	4 days.	Fluid drachms (?) Great relief, due probably to gradual escape of fluid into tissues around pericardium.	Hypertrophy of heart.	Letter from operator.

RESULTS OF THE OPERATION.

In preceding chapters I have considered the indications
for performing the operation of paracentesis of the pericardium, have discussed the various methods recommended, have given my views as to the claims of the
numerous points of puncture proposed, and have devoted
some space to discussing the objections urged against the
procedure. It now remains to study the table of operations placed at the beginning of this chapter, in order to
see what are the results that have been obtained, and that
we may expect to obtain, from the performance of paracentesis of the pericardium.

In the table there are recorded 60 cases of the operation.*   Of these, there were

Males   43
Females 12
Sex not mentioned 5

It may be interesting to study the relation of the operation and the ages of the patients so treated :

The cases under twenty years (inclusive) numbered   . 20
   "   over twenty years numbered 25
   "   whose age was not given numbered   . . 15

The greatest age at which tapping was done was sixty-
eight years, and the patient in this instance (No. 28) was
a female, who was operated upon twice. It is deemed
worthy of notice that this is apparently the only time that

---

* McCall Anderson recently aspirated the pericardium of a boy, aged
seventeen years.   Great improvement followed; but the report was made
just after the operation, and it is not possible to state the result.   See
Glasgow Medical Journal, September, 1879, p. 216.

the puncture was made on the right side of the sternum. It has previously been stated that this side has quite recently been suggested as the most available spot. The youngest patients were those of Roger (No. 35) and Villeneuve (No. 47), who were five years of age.

In all branches of medicine the main question that arises, when a line of action is advocated in therapeutics, is this : What is the ratio of recovery? This query I now attempt to answer :

The recoveries were	24
" deaths were	36
Total	60

It will be observed that I have included under the heading of fatal cases the patient operated on by Jowett (No. 4), where it is stated that there was a "hope of recovery." This is done, because I do not wish the statistics to give a more favorable view than is just. There are few probably who distrust statistics as much as I do, since so much depends on the accuracy and non-partisanship of the reporter. Especially is this the fact when a man reports a number of cases occurring in his own practice. That unknown quantity, the "personal equation" as to diagnostic skill, accuracy, and veracity, vitiates, to my mind, the great majority of such results. Hence I have endeavored in compiling these statistics to lean towards the side of the unbeliever, rather than to make too fair a showing for my side of the question.

Taking, then, the recoveries as twenty-four, and the deaths at thirty-six, we have 40 per cent. as the average of recovery, or 60 percentage of mortality. This average is certainly a good one, when the almost always fatal result of expectant treatment is remembered. If the fluid be not evacuated, the quantity increases until pressure upon, and maceration of, the heart, as well as the injurious tension to which the surrounding intra-thoracic structures are subjected, cause the death of the patient after most distressing symptoms, with five pints* or more of pus in the enormously distended sac.

The mortality after tracheotomy in croup in the St. Eugénie Hospital is, according to Barthez,† about 66⅔ per cent., and the number of cases that die after herniotomy is certainly great; yet these operations are accepted as justifiable. Why, then, should one hesitate to tap the pericardium in large effusions, when the fatality of let-alone treatment is fully recognized? The mortality of 60 per cent. is certainly better than that mentioned above as following tracheotomy. A man who would open a child's trachea in a trice for croup, would in many cases, I fear, let that child's father die with pericardial effusion, because he dare not tap the pericardium, and thus remove the agent preventing the proper oxygenation of blood as effectually as the membrane in the child's larynx. The mortality in paracentesis pericardii, given above, is inclusive of all cases found in the table, but very many of the patients had serious diseases complicating the pericardial effusion. Of the thirty-six who died, there were thirty-one who are known by us to have had other concomitant and often incurable disease. In the five remaining instances there was no other disease, or at least none mentioned. This would give the astonishing result of only five cases of death from uncomplicated pericardial effusion in a series of sixty cases of operation.

Let us look, however, at the results of the operation in recent times only, for the earlier cases cannot be scruti-

---

* Boston Medical and Surgical Journal, February, 1866, p. 29.

† Atkin's Practice of Medicine, 2d American ed., vol. ii. p. 998.

nized with as much thoroughness as desirable, and, moreover, do not give as minute details as I wish to obtain. To avoid the influence of such cases as Romero's, Kyber's, Aran's, and Béhier's, whose authenticity has been questioned and cannot be absolutely proved at this period, I shall take the cases that have occurred since 1860. In this number I include the fatal cases of Chaillou and Labric, whose dates I have not obtained, since it is probable that the operation was performed after 1860. The record will then stand as follows:

Since 1860 there have been .	.	10 recoveries.
" " " .	.	25 deaths.
Total	.	35 cases.

This gives a mortality of 71.42+ per cent. In the twenty-five instances where death occurred subsequent to the tapping, serious disease is stated to have existed in all the cases except three (Nos. 31, 34, and 38). In other words, out of the whole thirty-five cases operated upon, there were thirteen cases of pericardial effusion where other diseases did not seem to act as a complication, and of these ten recovered and three died. This would give a mortality of 23+ per centum.

By looking over the whole table it will be seen, as stated on page 79, that the average time of survival of those who are known to have lived beyond the first day is over twenty-seven days, which is equivalent to the assertion that those who are not actually moribund, and who survive the shock of operation, have a probable prolongation of life of nearly four weeks.

Thus, then, I have gleaned from the long list of cases the facts seeming to me most important, and by eliminating disturbing factors have presented a statement meant to be impartial. I am undoubtedly an advocate of the operation in selected cases, and may have unconsciously misrepresented the views of others, or unduly assumed that my side of the question was the correct one. If I have done so, I must regret it, for no scientific progress is to be expected until writers step outside of self and weigh pro's and con's with an assayer's balance.

Surely the record I have given supports the plea for the adoption of paracentesis of the pericardium into the family of accepted surgical procedures; surely it is noble to add three or four weeks to the life of a fellow-being. What care I if you call it a palliative operation, forsooth! Do we not excise carcinomatous breasts? Does not every one tap ascitic bellies, when cirrhotic liver exists, for palliation? Would you withhold opium from a groaning patient because it did not extirpate the disease? Who is to estimate the value of one week added to the life of a Cæsar? A perfect hip-joint is not expected after chronic coxitis, neither must a perfect heart be looked for after chronic pericarditis. Let the operation be palliative, if you will; but operate, and that before continuance of inflammation, maceration of the heart, and pressure of the distended sac have caused the tissues to assume pathological aspects of such a kind that a perfect return of function is impossible.

INDEX.

THE END.

www.ingramcontent.com/pod-product-compliance
Lightning Source LLC
Chambersburg PA
CBHW021946190326
41519CB00009B/1154